INTRODUCTION TO MATHEMATICAL FLUID DYNAMICS

RICHARD E. MEYER

DOVER PUBLICATIONS, INC.

NEW YORK

To H. Hopf
J. Ackeret
S. Goldstein
K. O. Friedrichs

and to the memory of
L. Prandtl

Published in Canada by General Publishing Company, Ltd., 30 Lesmill Road, Don Mills, Toronto, Ontario.
Published in the United Kingdom by Constable and Company, Ltd., 10 Orange Street, London WC2H 7EG.

This Dover edition, first published in 1982, is an unabridged and slightly corrected republication of the work originally published by Wiley-Interscience, a division of John Wiley & Sons, Inc., New York, in 1971.

Manufactured in the United States of America
Dover Publications, Inc.
180 Varick Street
New York, N.Y. 10014

Library of Congress Cataloging in Publication Data

Meyer, Richard E., 1919-
 Introduction to mathematical fluid dynamics.

 Reprint. Originally published: New York: Wiley-Interscience, 1971. (Pure and applied mathematics; v. 24). With slight corrections.
 Bibliography: p.
 Includes index.
 1. Fluid dynamics. I. Title. II. Series: Pure and applied mathematics (John Wiley & Sons); v. 24.
QA911.M48   1982   532'.05   82-7281
ISBN 0-486-61554-5 (pbk.)   AACR2

Contents

Preface

This book grew out of a semester course for beginning graduate students of mathematics, physics, engineering, and meteorology. Some of them had, and others had not, taken an undergraduate course in fluid dynamics. It became clear that they had a common interest in a certain core of knowledge on the subject, and that a good understanding of it would provide a very suitable basis for their diverse, specialized graduate courses on fluid mechanics and would raise their level of understanding in those courses quite drastically. This core of knowledge concerns the relation between inviscid and viscous fluids, and the bulk of this book is devoted to a discussion of that relation. The book is strictly an introduction, aiming to open the specialized literature for the reader, not to replace it. At the same time, an attempt is made to give the introduction at a level of sophistication appropriate to the subject, and to graduate students.

The concept of large Reynolds number is crucial to such a purpose, but does not, by itself, provide an introduction to theoretical fluid dynamics. A chapter on two topics in gas dynamics and a very brief sketch of some classical concepts on rotating fluids have therefore been added. Several other subjects should be touched upon in any introduction to fluid dynamics, but the cutoff is equally unsatisfactory wherever it comes.

In teaching a course of this type, I noted that the conventional, cursory treatment of kinematics injects a feeling of insecurity into the students' relation to fluid dynamics, to which later difficulties may be traced. Chapter 1 attempts to improve this situation. But in order to make a rapid start with the more important and exciting subject of dynamics, many readers may wish to begin with a rapid glance through this chapter and to return to it only later, as the need comes to be felt.

The somewhat axiomatical slant of Chapters 1 to 3 stems from the success of a pedagogical experiment in the classroom. For mathematics students such a treatment helps to dispel the all too common impression that the whole subject is built on a quicksand of assorted intuitions. For the students of physical or engineering sciences it helps to clear up common confusions between approximations of different types. Of course, the following account is not axiomatic; the "postulates" are used to illuminate the subject, not to deduce it.

Mathematicians will be even more disappointed by the lack of any attempt to prove an existence theorem or even to talk seriously about partial differential equations. There would seem to be room, however, for an introduction to what the whole subject is about, before the Navier-Stokes equations are tackled. In fact, many mathematicians will complain that the book contains hardly any mathematics; and many theoretical fluid dynamicists, that the book is far too abstract. This would indicate a gap which should be filled.

To make the book widely accessible, only a modest mathematical preparation is presupposed—essentially, a thorough understanding of calculus and vector analysis, including the Divergence theorem. More elaborate technical tools are explained where they intrude, unless they arise only in a peripheral way, so that the reader to whom they are unfamiliar is not put at a serious disadvantage. The literature references are restricted to an indication of texts in which the reader can look for help on such occasions, if he wishes to persist; no attempt at a proper bibliography has been made.

Madison, Wisconsin
August 1970

R. E. MEYER

Acknowledgments

The author is indebted to many friends and colleagues who stimulated or transmitted ideas which make up this book. In particular, E. R. Fadell, K. O. Friedrichs, S. Y. Husseini, C. R. Illingworth, A. Kadish, J. J. Mahony, W. H. Reid, M. C. Shen, J. J. Stoker, and F. Ursell read parts of the manuscript and gave helpful advice.

A massive debt is owed to the United States Air Force Office of Scientific Research for Grant AFOSR-1248-67 to support the writing of this book; and a similar debt for support is owed to the Mathematics Research Center, United States Army. Further support for research incorporated here was provided by the National Science Foundation and by the University of Wisconsin.

Permission to reproduce material from the following sources is acknowledged: from L. Prandtl, *The Essentials of Fluid Dynamics*, Blackie, London, 1952, in respect of Figs. 4.2, 6.9, 6.10, 6.12, 8.2, 9.3, 13.2, and 30.1; from A. Betz, *Konforme Abbildung*, 2nd ed., Springer, Heidelberg, 1964, in respect of Figs. 4.1, 6.1, 6.2, 6.3, 6.4, 6.6, 6.7, and 6.8; from Sir H. Lamb, *Hydrodynamics*, 6th ed., Cambridge University Press, 1932, in respect of Fig. 27.4; from O. G. Tietjens, *Fundamentals of Hydro- and Aeromechanics*, McGraw-Hill, New York, 1934, in respect of Figs. 6.5 and 6.14; from V. L. Streeter (Ed.), *Handbook of Fluid Dynamics*, McGraw-Hill, New York, 1961, in respect of Fig. 22.2; and from H. R. Byers, *General Meteorology*, 3rd ed., McGraw-Hill, New York, 1959, in respect of Fig. 30.2.

The half-tone plate on p. 111 (Fig. 23.1) has been contributed by the National Physical Laboratory, Teddington, Mddx, through the kind offices of Dr. R. J. North of the Aerodynamics Division; the photograph is Crown Copyright Reserved.

CHAPTER 1

Kinematics

1. Introduction

Of the two possible didactic approaches to a physical subject, one proceeds from general definitions to their implications, leaving aside the problem of deciding which definitions are relevant to a particular, real situation. The other proceeds from specific situations and examples to definitions there relevant leaving the general principles of the subject to emerge only by and by. Fluid dynamics is too varied and subtle a subject for the first approach, but the second is too hard on the student, who has to persevere through an inordinate volume of work before the subject takes over-all shape. Like most other textbooks, this work therefore adopts largely the first approach.

One of its notable shortcomings is that the physical scales remain vague during the discussion of the general laws of fluid dynamics, because these laws can be made nondimensional only by reference to specific situations. An important part of the nature of the subject will therefore become manifest only in the second half of the book. For instance, a practical criterion for the applicability of continuum fluid dynamics—to which this book is restricted—will not be stated before Section 23. This long postponement arises from the subtlety of the considerations involved in any more than superficial discussion of the general definitions; Chapters 2 and 3 are needed, in part, to explain why nothing more simple-minded is adequate.

The definition of fluid motion will be presented in four stages (Chapters 1, 2, 3, and 6) of which the first two concern a very general continuum concept of fluid, which is necessarily incomplete. Of course, any model of a fluid that can be discussed mathematically is an idealization, and therefore incomplete. The term "ideal fluid," however, has acquired a specific, technical meaning (Section 13) denoting a radically simplified model. By antithesis, the Newtonian fluid is often called "real" because it is a highly realistic model of many common fluids under a wide range of circumstances. It is the fluid model primarily considered in this book, and the conventional term "real fluid" will

1

be used for it on occasion, especially for comparison with the "ideal fluid" model in Chapter 2.

The physical definition of the (incompressible) Newtonian fluid will be given in the form of ten "postulates." Their purpose is to differentiate this definition from the manifold approximations (often most conveniently introduced as assumptions motivated intuitively) which are needed as aids to the description of fluid motion. Such differentiation is helpful in the earlier stages of a study of fluid dynamics, but the categorical appearance of the postulates is not intended to hide the fact that they become properly fruitful only when complemented by the nondimensional formulation of definite problems (Chapters 4 and 5). The further definitions needed for the compressible fluid are introduced more conventionally in Chapter 6.

Appendix 1

Notation. Equations, figures, problems, and statements cast for ease of reference in the form of theorems, lemmas, or corollaries are numbered by section. Appendices are numbered by the sections to which they are attached. Numbers in brackets [ ] refer to the bibliography at the end of the book, which aims only to serve the convenience of readers wishing to check on arguments for which there is no space in this book.

Bold letters denote vectors in a three-dimensional Euclidean space E^3. Their components with respect to some particular Cartesian coordinate system are usually denoted by subscripts, for example, $\mathbf{x} = \{x_1, x_2, x_3\}$. On occasion, the alternative notation $\mathbf{x} = \{x, y, z\}$ and $\mathbf{v} = \{u, v, w\}$ is employed for convenience.

Tensors will play a minor role, and the reader not familiar with them will find it sufficient to interprete them as matrices with nine elements t_{ij}, $i, j = 1, 2, 3$ (the "components"), dependent on the Cartesian coordinate system used in such a way that $\sum_{j=1}^{3} t_{ij} b_j$, $i = 1, 2$, or 3, are the components of a vector whenever b_j, $j = 1, 2, 3$, are the components of a vector. No special symbol for tensors will be introduced, since no serious confusion will arise in the following from letting t_{ij} denote the tensor as well as its individual components. A particular tensor occurring frequently is Kronecker's symbol δ_{ij}, defined by $\delta_{ij} = 0$ for $i \neq j$ and $\delta_{ij} = 1$ for $i = j$.

The summation convention will always be used unless the contrary is stated explicitly; it is to sum automatically over any repeated subscripts which are not separated by a $+$, $-$, or $=$ sign; for instance, $\delta_{ii} = 3$.

Logic Notation. It will be convenient and will help to clarify the text to make fairly frequent use of the following symbols.

$a \in B$ for the object a is an element of, or belongs to, the set B;

$S_1 \subset S_2$

or $S_2 \supset S_1$ for S_1 is a subset of S_2;

$\{a | \beta\}$ for the set of all objects a to which the statement β applies;

$S_1 \cup S_2$ for the union of the sets S_1 and S_2, i.e., the set $\{a | a \in S_1 \text{ or } a \in S_2\}$;

$S_1 \cap S_2$ for the intersection of the sets S_1 and S_2, i.e., the set $\{a |$ both $a \in S_1$ and $a \in S_2\}$.

Point Set Notation. For the purposes of this book, adequate definitions of the terminology used in Section 2 and, with varying frequency, also in later sections are as follows. If x_0 denotes the position vector of any chosen point P in E^3, the set of points with position vectors x such that $|x - x_0| < \varepsilon$ is called a *neighborhood* of P, if $\varepsilon > 0$. A point-set S is *open* means that $P \in S$ implies P has a neighborhood $\subset S$. S is *bounded* means that a spherical ball Σ of finite radius can be found such that $S \subset \Sigma$. Q is a *limit point* of S means that every neighborhood of Q contains a point $\in S$ distinct from Q. A set R is *closed* means that all its limit points also belong to R. On the other hand, the *boundary* ∂S of an open set S is the set of all those limit points of S which do not belong to S. The *closure* $\bar{S}$ of an open set S is $S \cup \partial S$. A standard example of an open point set is $S = \{x | 0 < |x| < 1\}$; its boundary ∂S consists of the sphere $|x| = 1$ and the point $x = 0$; its closure is $\bar{S} = \{x | |x| \leq 1\}$. The distinction between points and their position vectors will usually be omitted.

Order Symbol. The asymptotic notation $\psi(y, z) = O(\chi(y, z))$ as $z \to \infty$ (or $\to 0$) means that there are positive numbers D and M such that $z > D$ (or $< D$) implies $|\psi(y, z)| < M|\chi(y, z)|$ for every fixed y in the domain of ψ or in some other, clearly specified domain. An o in the place of O means that the statement holds even for arbitrarily small M. Another useful notation is $\psi \sim \chi$, meaning that $|\psi - \chi| \to 0$ as $z \to \infty$ (or 0) for every fixed y. The word "fixed" here is meant to recall that such an asymptotic statement refers to a single limit process (with respect to z) in which ψ and χ are considered at a value of y independent of z.

2. Lagrangian Description

To cast our intuitive feeling for the motion of fluids (i.e., liquids and gases) into a definition on which mathematical argument can be based, is a more complicated task than might be thought at first. The attempt to be made now will not be completed until Section 17, even for the simplest fluids of our everyday experience. To make a start, the notion of some *identifiable* entity which might be called a "body of fluid" would appear to be near the core of our intuitive ideas. It is supported by the simple experiences of watching a puff of smoke in an air stream, or a drop of ink in a liquid stream, as they float along. A drop of ink, for instance, which is of the same density as the surrounding transparent liquid, may be plausibly considered a part of the liquid distinguished from the rest only by its color, but not by any dynamically relevant properties. If the ink is not soluble in the liquid and diffuses only slowly, and if the motion is not too involved, there is no difficulty in maintaining the visual identification of the colored liquid, as it moves and deforms, for a considerable time.

This leads naturally to the idea of associating fluid motion with a geometrical transformation represented by a function $x = x(a, t)$ giving the position vectors x at various times t of the "bit" or "element" of fluid identified by the label a. A description of fluid motion thus based on identifying the individual bits of fluid is called Lagrangian. One simple way of labeling is to let a denote the position vector at some chosen time, which may then be called $t = 0$.

While this suffices for some purposes, the notion of "bit" or "element" is too confused for others. The underlying idea is expressed more clearly by the following

Definition. Let Ω_0 be any open, bounded point-set in E^3 "occupied by fluids" at time $t = 0$. "Fluid motion" shall mean a transformation H_t on the closure $\overline{\Omega_0}$ into E^3 such that the point set $H_t\Omega_0$ is that occupied by "the same fluid" at time t.

Readers unfamiliar with some of these terms will find definitions in Appendix 1. The notion that the set occupied by a fluid should be regarded as open is both mathematically necessary and suggested by physical intuition; it is difficult to think of a fluid as occupying a closed set, for that would imply, for instance, in the case of a solid spherical shell filled with water and surrounded by air, that the solid is restricted to an open point set. On the other hand, it is plausible that the set of mappings H_t describing how the body of fluid moves from Ω_0 into the set $H_t\Omega_0$ should be such that some statements are also possible regarding the mapping of the boundary points of Ω_0. That is expressed in the definition regarding H_t as defined on the closure $\overline{\Omega_0}$.

The notion of identification may not at first appear natural for the description of processes such as the mixing of two chemically distinct fluids by diffusion, but it is actually quite convenient to regard the mixture itself as the fluid and the concentration of one chemical in the other as one of the fluid properties varying with **a** and t.

Any definition of fluid motion in terms of a geometrical transformation requires specification of the times t for which the transformation is defined, and most accounts of fluid dynamics tend to give the impression that H_t is defined on $\overline{\Omega_0}$ for all real t. Experience does not really support this. The difficulty becomes clear in any attempt to identify a blob of ink in a turbulent river for more than a few moments. The motion is too involved, and the blob of ink soon stretches into a ribbon so entwined and entangled with clear liquid that visual distinction between clear and colored liquid becomes prohibitively difficult. The study of such motions has raised serious doubt whether identification can ever be maintained indefinitely. On the other hand, no circumstances are known in which it cannot be maintained over a sufficiently short time, and experience therefore suggests

Postulate I. H_t is defined on $\overline{\Omega_0}$ over a time interval (T_1, T_2) such that $T_1 < 0 < T_2$.

The first task, then, is to explore the nature of fluid motion during such an interval, and that provides ample scope for a book such as this. The further discussion will therefore be based on the assumption that $|T_1|$ and T_2 are as large as may be required.

The notion of identification is closely related to the intuitive idea that two different bodies of fluid cannot simultaneously occupy the same portion of space. To express this with respect to the concept of fluid motion just defined requires

Postulate II. H_t has an inverse on $H_t\Omega_0$.

Thus, if H_t is represented by the function $\mathbf{x} = H_t\mathbf{a} = \mathbf{x}(\mathbf{a}, t)$ on $\overline{\Omega_0}$, there exists an inverse function, $\mathbf{a} = \mathbf{a}(\mathbf{x}, t)$, mapping $H_t\Omega_0$ onto Ω_0. However (Appendix 2), a single-valued inverse cannot always be defined on the closure of $H_t\Omega_0$.

Most general accounts of fluid dynamics do not cover processes involving drop formation and evaporation, and the same limitation will be adopted here. Common experience then indicates clearly that fluid motion is a continuous process in the sense of

Postulate III. $\mathbf{x} = \mathbf{x}(\mathbf{a}, t)$ is continuous with respect to $\mathbf{a}$ and t on its domain of definition, and its inverse is similarly continuous.

This expresses that it is impossible to break or cut a fluid—one of the basic points distinguishing it from a solid. The intuitive idea of a solid involves limitations on the possible extent (or rate) of continuous deformation. By contrast, a body of fluid is intuitively felt to be capable of changing shape continuously to an arbitrary extent. Moreover, an attempt to cut fluid results only in the fluid wrapping itself around the knife. A simple experiment is to move a knife through a liquid in which an ink filament has been laid across the path of the knife; it will be observed that the filament is stretched, but not severed, as indicated schematically in Figs. 2.1 and 2.2.

Observe the implications of Postulate III that $H_t\Omega_0$ must, like Ω_0, be a bounded set. Our basic definitions and postulates will be formulated only with reference to such point sets, since we cannot have direct experience or intuition with respect to any others. Of course, extrapolation to unbounded sets will often be mathematically convenient.

It follows [e.g., 1, Vol. 2, p. 97] from Postulates II and III that the set $\Omega_t \equiv H_t\Omega_0$ is also open, and that its boundary is

$$(2.1) \qquad \partial\Omega_t = H_t\,\partial\Omega_0.$$

This is sometimes expressed by the phrase "the bounding surface of a body of fluid always consists of the same fluid particles" (see Appendix 2).

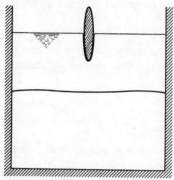

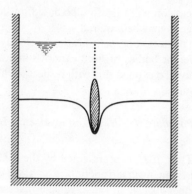

Fig. 2.1. Schematic illustration of an early stage in the process of moving a knifelike body downward through a liquid in a tank.

Fig. 2.2. Schematic illustration of a later stage in the process of moving a knifelike body downward through a liquid in a tank.

All experimental experience indicates that Postulate III does not fully reflect the continuity of fluid motion and should be amplied by

Postulate IV. The velocity $\mathbf{v} \equiv \dfrac{\partial}{\partial t}\,\mathbf{x}(\mathbf{a}, t)$ is defined on the domain of $\mathbf{x}(\mathbf{a}, t)$.

Appendix 2

A brief comment on a less than obvious implication of Postulate III may help to understand it better. If a knife is moved through a body of fluid, will the fluid close up behind the knife? In seeming contradiction to observation, the continuity of $\mathbf{x}(\mathbf{a}, t)$ implies a less than affirmative answer! Since H_t is continuous on its full domain $\overline{\Omega_0}$, the boundary $\partial\Omega_0$ cannot be severed by the motion, however much it be stretched, and a subset S of $H_t\,\partial\Omega_0$ must therefore be left behind in the fluid, as indicated schematically in Fig. 2.2 by a dotted line (the actual position of S is, of course, not predicted by this mathematical conclusion). Even though usually invisible, S must, by (2.1), be regarded as part of the boundary of the fluid domain Ω_t. The reader may note in passing that S provides an example of a subset of $\partial\Omega_t$ on which the inverse mapping H_t^{-1} is not single-valued. In many respects such a boundary set completely embedded in fluid is irrelevant and may be disregarded. Strictly speaking, however, that amounts to disregarding the distinction between two different fluid motions. One is the original fluid motion defined on Ω_0 (and thence on $\overline{\Omega_0}$, by continuous extension). The other is a new fluid motion defined on the open set which is, not $H_t\Omega_0$, but $(H_t\Omega_0) \cup S$. The second fluid motion is therefore not strictly a continuation of the original one. This hair-splitting distinction can have practical importance; for instance, it is part of the key to understanding the generation of aerodynamic lift (Appendix 14). To put it differently, Postulate III is seen to contain the seeds of the wake concept.

3. Eulerian Description

There is usually no need for an explicit consideration of the function $x(a, t)$ representing a fluid motion. A description by means of the velocity field v, considered as a function of x and t, suffices for virtually all practical purposes. Such a representation of fluid motion, akin to a motion picture of the *velocity field*, is called Eulerian and is used with only rare exceptions by fluid dynamicists.

Accordingly, unless an explicit statement is made to the contrary, all functions introduced below are understood to be defined as functions of x and t on a *fluid domain*—meaning a family of open bounded point sets $\Omega_t = H_t\Omega_0$ occupied by fluid during an open interval of t. Closure, in this connection, will still refer to $\overline{\Omega_t}$, and "fluid domain at time t" will denote Ω_t for fixed t. To characterize the continuity of such functions, $C^n(\Omega_\tau)$ will be used to denote the class of functions possessing continuous partial derivatives of order $\leq n$ with respect to both x and t on Ω_t during a time interval including τ.

It will, in fact, help to agree on some convention regarding the general function class to be envisaged for the velocity field. In principle that function class is determined by the postulates defining the fluid, but the task of deducing it in general from the postulates has not yet been accomplished. A number of exact solutions of the postulates are known (of which those of Problems 19.1 to 19.4 and Section 37 are rather typical), and their velocity fields are in $C^\infty(\Omega_t)$. This smoothness is traceable to the constitutive equation (Section 17) which describes a mechanism of diffusion of momentum and thereby indicates a probabilistic substratum in the foundations of fluid dynamics.

There is, in fact, overwhelming evidence supporting the view that gases and liquids consist of very large numbers of very small molecules interacting by collisions so that their joint behavior is describable statistically. The fluid velocity v, and the density and stress tensor to be introduced in Sections 4 and 10, are to be interpreted as appropriate averages of mechanical properties of the molecules. Continuum fluid dynamics—to which this book is restricted—is then defined as the study of a limit in which the number of molecules in "unit volume" tends to infinity, while the typical time and distance between successive collisions for any individual molecule tend to zero by comparison with "unit time" and "unit length." (But it is not at all implied that a body of fluid consists of the same molecules at all times. The term "particle" has different meanings in continuum mechanics and in kinetic theory, and is partially misleading in both fields.) There are strong grounds for assuming that in this limit the averages tend to smooth functions of both x and t. The precise degree of smoothness is not of decisive mathematical relevance in this book; physically it raises the question—fruitless in our context—of where the boundary between kinetic theory and continuum fluid dynamics should be drawn. Accordingly it will be best to adopt here the view that real velocity fields satisfy the

Smoothness convention. The velocity **v** (and the density and stress tensor to be introduced later) possesses partial derivatives of all orders with respect to **x** and t on the closure of the fluid domain.

As a rule, arguments in the following will be based on this convention, even when a weaker continuity assumption would suffice. Some exceptions will be made, however, because it is often convenient to approximate real velocity fields by limiting fields which are not smooth. This asymptotic technique can contribute much to the clarity of the mathematical description and has proved very fruitful in fluid dynamics. Indeed, a major aim of Chapters 2 and 4 will be the elucidation of some ways in which the concept of inviscid fluid leads to such limiting fluid motions. It will thus be helpful to have a criterion distinguishing arguments concerning the velocity, density, and stress fields of the basic continuum fluid ("real fluid") from arguments concerning limiting motions intended only to represent asymptotic descriptions of real motions. The smoothness convention serves this purpose, just as the postulates serve to distinguish the basic fluid model from auxiliary, asymptotic models.

The Lagrangian mapping H_t representing real fluid motion must then also be smooth. Indeed, by Postulate IV, $\mathbf{x}(\mathbf{a}, t)$ is the solution of the differential equation

$$\partial \mathbf{x}/\partial t = \mathbf{v}(\mathbf{x}, t)$$

with initial condition $\mathbf{x}(\mathbf{a}, 0) = \mathbf{a}$ on $\overline{\Omega_0}$. By the smoothness convention, $\mathbf{v} \in C^\infty(\overline{H_t \Omega_0})$ for all times covered by Postulate I, and as noted in Section 2, any other times are ignored in this book. The fundamental theorem of ordinary differential equations [e.g., 2] therefore assures the existence of a unique solution $\mathbf{x}(\mathbf{a}, t)$ with time derivatives of all orders for each $\mathbf{a} \in \overline{\Omega_0}$. This solution depends continuously on the initial condition [2] in the sense that, if a set of initial conditions $\mathbf{a}(\sigma)$ depends on one or more parameters σ, then the solution set $\mathbf{x}(\mathbf{a}(\sigma), t)$ possesses as many continuous derivatives with respect to σ as $\mathbf{a}(\sigma)$ possesses.

Since the basis of fluid dynamics is part Eulerian and part Lagrangian, a short dictionary is required for translating from one representation to the other. If any quantity has the Eulerian representation $f(\mathbf{x}, t)$, its Lagrangian representation is $g(\mathbf{a}, t) \equiv f(\mathbf{x}(\mathbf{a}, t), t)$, and its *material* or *convective derivative* is defined as

$$\frac{Df}{Dt} \equiv \frac{\partial}{\partial t} g(\mathbf{a}, t).$$

This is the rate of change with time of the property f of the fluid "bit" or "element" which passes through the point $\mathbf{x}$ at the time t. It differs generally from $\partial f/\partial t$; in fact,

$$(3.1) \qquad \frac{Df}{Dt} \equiv \frac{\partial g}{\partial t} = \frac{\partial f}{\partial t} + \frac{\partial f}{\partial x_k}\frac{\partial x_k}{\partial t} = \frac{\partial f}{\partial t} + v_k\frac{\partial f}{\partial x_k} = \left(\frac{\partial}{\partial t} + \mathbf{v}\cdot\text{grad}\right)f,$$

by Postulate IV. (As set out in Appendix 1, the subscripts denote Cartesian components, and the summation convention is understood.) As an example, the *fluid acceleration* is defined as

$$(3.2) \qquad D\mathbf{v}/Dt = \partial\mathbf{v}/\partial t + (\mathbf{v}\cdot\text{grad})\mathbf{v}.$$

Another important example occurs in the description of a bounding surface of the fluid. The word "surface" has many meanings, and in addition to the intuitive name surface for certain types of point sets, the term *regular surface* will be used to denote a simple, piecewise smooth, orientable surface in E^3 to which the Divergence theorem is applicable (a more explicit definition will be found in Appendix 3). The phrase *surface moving with the fluid* will denote a family of surfaces $S(t) = H_t S_0$, where S_0 is a fixed surface in Ω_0; but in keeping with usual practice, $S(t)$ will be abbreviated to S. (The smoothness of H_t can be used to show that $H_t S_0$ is a regular surface if S_0 is regular.) If a surface S moving with the fluid has the equation $F(\mathbf{x}, t) = 0$, then it follows that $F(\mathbf{x}, t) = G(\mathbf{a}, t)$ such that

$$(3.3) \qquad \partial G/\partial t = DF/Dt = 0 \qquad \text{on } S.$$

It is often convenient to express this statement in terms of the velocity $\mathbf{V}$ of the surface normal to itself. To this end, note that $F(\mathbf{x}, t) = 0$ defines $\mathbf{x}$ as a function of t such that

$$dF/dt = (\partial F/\partial x_i)\,dx_i/dt + \partial F/\partial t = 0$$

whenever $\mathbf{x}$ and t satisfy $F(\mathbf{x}, t) = 0$. This relation fails to define dx/dt because a set of points forming a geometrical surface S admits many mappings into itself. The ambiguity disappears, however, when only the component of dx/dt normal to S is considered, and the normal velocity of a regular surface $F(\mathbf{x}, t) = 0$ is therefore

$$\mathbf{V} = (\mathbf{n}\cdot dx/dt)\mathbf{n} = -\mathbf{n}(\partial F/\partial t)/|\text{grad } F|$$

in the sense of the unit normal $\mathbf{n} = (\text{grad } F)/|\text{grad } F|$ (which is a piecewise continuous function of position on a regular surface). It now follows from (3.1) that (3.3) is equivalent to

$$(3.4) \qquad (\mathbf{v} - \mathbf{V})\cdot\mathbf{n} = 0;$$

that is, *the normal component of the relative fluid velocity is zero at every point of a regular surface moving with the fluid.*

The boundary of an individual body of fluid is, by the definition of fluid motion, the set $H_t\ \partial\Omega_0$, where Ω_0 is the set occupied by this body of fluid in the Lagrangian a-space. It follows that (3.3) applies to such a boundary. Moreover, $H_t\ \partial\Omega_0$ is either regular or a union of point sets which are regular surfaces and other point sets which are not, and the latter are definable as limits of point sets which are regular surfaces moving with the fluid, so that (3.4) still applies. Accordingly (3.3) and (3.4) are seen to be the *basic kinematic boundary condition* of the fluid.

Convection theorem 3.1. If Ω_t is a fluid domain and if $f(\mathbf{x}, t) \in C^1(\overline{\Omega}_t)$, then

$$(3.5) \qquad \frac{D}{Dt}\int_{\Omega_t} f\, dV = \int_{\Omega_t} \left(\frac{Df}{Dt} + f\operatorname{div} \mathbf{v}\right) dV,$$

where dV denotes the volume element.

Corollary 3. With Ω_t and f as in the convection theorem, let Ω_1 be the fixed set in E^3 which coincides with Ω_t at $t = t_1$ and assume $\partial\Omega_1$ to be a regular surface. Then at the (arbitrary) time t_1

$$(3.6) \qquad \frac{D}{Dt}\int_{\Omega_t} f\, dV = \frac{\partial}{\partial t}\int_{\Omega_1} f\, dV + \int_{\partial\Omega_1} f\mathbf{v}\cdot\mathbf{n}\, dS,$$

where $\mathbf{n}$ is the unit outward normal, and dS the surface element, on $\partial\Omega_1$.

A proof may be found in Appendix 3. The corollary clearly relates Lagrangian and Eulerian concepts. The left-hand side of (3.6) is the rate of change of the f-content of the *fixed body of fluid* occupying the domain $\Omega_1 \subset E^3$ at time t_1. The first term on the right-hand side is the rate of change of the f-content of this *fixed spatial domain*. And the last term is the rate of outflow of f through the fixed boundary of Ω_1 ("flux of f through $\partial\Omega_1$").

Appendix 3

Regular Surfaces. Kellogg [3] defines the term regular surface as follows. A regular region R is understood to be a closed and bounded set $\subset E^2$ such that its boundary ∂R is a simple circuit (see Section 5). A regular surface element S is defined as a point set $\subset E^3$ admitting with respect to some Cartesian coordinate system a representation $x_3 = f(x_1, x_2) \in C^1(R)$, where R is a regular region in the x_1, x_2-plane. Restriction of f to ∂R yields the boundary ∂S, which is also a simple circuit, and an edge of S is defined as a subset of ∂S on which $\mathbf{x}$ is a continuously differentiable function of the arc length. A vertex of S is a point common to two edges of S which is not an interior point of either.

A regular surface is defined as the union of a finite number of regular surface elements S_j such that, for all distinct i, k, m,
1) $S_i \cap S_k$ is one edge for both, or one vertex for both, or empty,
2) $S_i \cap S_k \cap S_m$ is a set of vertices for each, or empty,

3) S_i and S_k are the first and last of a chain of surface elements S_j, $i \leq j \leq k$, such that $S_j \cap S_{j+1}$ is an edge for S_j and for S_{j+1},

4) if the S_j for $i \leq j \leq k$ have a common vertex, they form a chain such that $S_j \cap S_{j+1}$ is an edge terminating at this vertex.

The definition of regular surface adopted in this book differs from this in two respects. First, the surface is required to be orientable, that is, to possess two unambiguously distinguishable sides (in contrast to Moebius' band, e.g.). Second, the union of a finite number of disjoint regular surfaces in Kellogg's sense is still called a regular surface.

Proof of the convection theorem. By hypothesis, $\Omega_t = H_t\Omega_0$, where Ω_0 is a bounded fixed set in a-space, and therefore

$$\int_{\Omega_t} f \, dV = \int_{\Omega_0} g(\mathbf{a}, t) \mathbb{J}(\mathbf{a}, t) \, dV_0,$$

where $g(\mathbf{a}, t) = f(\mathbf{x}(\mathbf{a}, t), t)$ and the Jacobian $\mathbb{J} = \det(\partial x_i/\partial a_k)$. Thus

$$\frac{D}{Dt} \int_{\Omega_t} f \, dV = \frac{\partial}{\partial t} \int_{\Omega_0} g\mathbb{J} \, dV_0 = \int_{\Omega_0} \frac{\partial}{\partial t} (g\mathbb{J}) \, dV_0,$$

the interchange being justified by the continuity of $\partial(g\mathbb{J})/\partial t$ following from the smoothness convention. But $\partial(g\mathbb{J})/\partial t = \mathbb{J} \, Df/Dt + f \, \partial\mathbb{J}/\partial t$ and the theorem follows, if $\partial\mathbb{J}/\partial t = \mathbb{J} \operatorname{div} \mathbf{v}$. Now the Jacobian determinant may be expanded by rows as $\mathbb{J} \, \delta_{ik} = (\partial x_i/\partial a_n)A_{nk}$, where δ_{ik} is Kronecker's symbol. To avoid differentiating the cofactors A_{nk}, let us momentarily denote the i-th row vector of $\mathbb{J}$ by $\mathbf{r}_i$, its components by $(\mathbf{r}_i)_n$, and the differential operator by ∂. Since the determinant is linear in its rows, its derivative is the sum of determinants,

$$\partial \det(\mathbf{r}_1, \mathbf{r}_2, \mathbf{r}_3) = \det(\partial\mathbf{r}_1, \mathbf{r}_2, \mathbf{r}_3) + \det(\mathbf{r}_1, \partial\mathbf{r}_2, \mathbf{r}_3) + \det(\mathbf{r}_1, \mathbf{r}_2, \partial\mathbf{r}_3)$$
$$= A_{n1} \, \partial(\mathbf{r}_1)_n + A_{n2} \, \partial(\mathbf{r}_2)_n + A_{n3} \, \partial(\mathbf{r}_3)_n,$$

so that

$$\frac{\partial\mathbb{J}}{\partial t} = A_{nk} \frac{\partial}{\partial t}\left(\frac{\partial x_k}{\partial a_n}\right) = A_{nk} \frac{\partial v_k}{\partial a_n} = A_{nk} \frac{\partial v_k}{\partial x_l} \frac{\partial x_l}{\partial a_n} = \mathbb{J} \, \delta_{lk} \frac{\partial v_k}{\partial x_l}.$$

To prove the corollary, it suffices to note that

$$Df/Dt + f \operatorname{div} \mathbf{v} = \partial f/\partial t + \operatorname{div}(f\mathbf{v})$$

by (3.1), that $\int \partial/\partial t = \partial \int/\partial t$ as in the theorem, and that

$$\int_{\Omega_1} \operatorname{div}(f\mathbf{v}) \, dV = \int_{\partial\Omega_1} (f\mathbf{v})\cdot\mathbf{n} \, dS$$

by the Divergence theorem.

4. Conservation of Mass

Postulate V. A function $\rho(\mathbf{x}, t)$ is defined on the closure of any fluid domain $\Omega_t = H_t\Omega_0$ so that for all t

$$\int_{\Omega_t} \rho \, dV = m(\Omega_0) > 0.$$

This positive invariant of the fluid motion is called the mass of the fluid in Ω_t, and ρ, its *density*. Of course, the postulate is understood to refer not only to the whole domain of a body of fluid but also to every subdomain.

It follows from (3.5) and (3.6) that

(4.1) $$\int_{\Omega_t} (D\rho/Dt + \rho \operatorname{div} \mathbf{v}) \, dV = 0$$

for every fluid domain Ω_t, and

(4.2) $$\frac{\partial}{\partial t} \int_{\Omega_1} \rho \, dV = -\int_{\partial\Omega_1} \rho \mathbf{v} \cdot \mathbf{n} \, dS$$

for every fixed domain Ω_1 (with regular boundary surface $\partial\Omega_1$) occupied by fluid during a time interval, and therefore

(4.3) $$D\rho/Dt + \rho \operatorname{div} \mathbf{v} = \partial\rho/\partial t + \operatorname{div}(\rho\mathbf{v}) = 0$$

everywhere in the fluid. (But note that the invariant m is directly observable, whereas (4.3) is only inferred by the smoothness convention, and applies for limiting motions only to sets on which ρ and $\mathbf{v} \in C^1$.) Equation (4.3), which is usually called the "equation of continuity," and the convection theorem imply

Corollary 4.1.   If $f(\mathbf{x}, t) \in C^1(\overline{\Omega_t})$, then

(4.4) $$\frac{D}{Dt} \int_{\Omega_t} \rho f \, dV = \int_{\Omega_t} \rho \frac{Df}{Dt} \, dV.$$

It is convenient here to introduce some terms in frequent use. The name **incompressible** motion denotes a fluid motion such that $\rho(\mathbf{x}, t) \equiv \text{const.}$ In fluid dynamics, incompressibility thus refers not to a property of the fluid but to a property of the representation by which its real motion is approximated. Some brief and partial indications of circumstances under which this approximation is justified will be given in Chapter 6.

A less restrictive definition, making incompressibility equivalent to the statement that $\rho(\mathbf{x}(\mathbf{a}, t), t)$ is a function only of the Lagrangian label $\mathbf{a}$, independent of t, is used for the study of "stratified fluids" in geophysics [e.g., 4]. There will not be room below for a discussion of that subject, and in the absence of an explicit statement to the contrary, the stricter definition will always be understood (more precisely, the almost universal custom will be followed of using "incompressible" loosely in the place of the terms "of constant density" or "isopycnic"). With either definition $D\rho/Dt \equiv 0$, so that (4.3) implies

(4.5) $$\operatorname{div} \mathbf{v} = 0.$$

Steady motion or *steady flow* is defined as motion such that ρ and $\mathbf{v}$ are independent of time. The visualization of such a motion may often be aided by tracing the *streamlines*, defined as the trajectories of the velocity field, i.e., the solution curves of $d\mathbf{x}/ds = \mathbf{v}/|\mathbf{v}|$, where s denotes arc length (Figs. 4.1, 2).

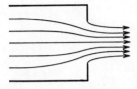

Fig. 4.1. Streamlines of a two-dimensional potential flow through an orifice.

By Postulate IV and the smoothness convention [2] precisely one such curve passes through each point of the fluid domain at which $\mathbf{v} \neq 0$. (Points at which $\mathbf{v} = 0$ are called *stagnation points*.) A more general concept, available

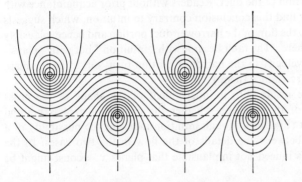

Fig. 4.2. Streamlines of the potential flow of a double row of vortices.

also in unsteady motion, is that of *particle path*, defined as a curve $\mathbf{x}(\mathbf{a}, t)$ for fixed $\mathbf{a}$. In steady flow the particle paths coincide with the streamlines by Postulate IV.

As an illustration of these concepts, consider steady flow in a duct the cross-sectional area of which varies with distance along the duct (Fig. 4.3); the duct walls are understood to be solid and impermeable, so that $\mathbf{v} \cdot \mathbf{n} = 0$ on them, by (3.4). Let the duct axis be straight and measure x along it in the general sense of the flow. Let A_i denote the respective cross-sectional areas of the duct in arbitrarily chosen planes $x = x_i$, $i = 1, 2$ (Fig. 4.3), and let u denote the x-component of the velocity. Then Postulate V for the fluid domain Ω bounded by the duct wall and the planes $x = \text{const} = x_i$ implies

$$\text{(4.6)} \qquad \rho_2 u_2 A_2 = \rho_1 u_1 A_1$$

by (4.2), where $\rho_i u_i$ stands for the average of ρu over the cross-section. In incompressible flow, in particular,

$$\text{(4.7)} \qquad u_2/u_1 = A_1/A_2,$$

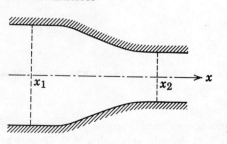

Fig. 4.3.

so that the average over the cross-section of the velocity u is seen to be larger in the narrower portion of the duct. Readers without prior acquaintance with fluid dynamics may find this conclusion contrary to intuition, which suggests greater resistance to the flow in the narrower duct portion and hence, plausibly a smaller velocity there. (The fault lies not with "intuition" but with "plausibility." The larger velocity in the narrower portion turns out to be the main cause of the larger resistance there.)

Observe that the result (4.6) is independent of the shape of the cross section, and even of the question whether the change of area is gradual (Fig. 4.3) or abrupt (Fig. 4.4), though intuition indicates the possibility of a marked difference between the details of the flow in the two cases. In the case of Fig. 4.4 it is indeed not implausible that planes $x = $ const might be

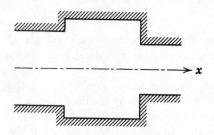

Fig. 4.4.

found in which the x-component of velocity has different signs at different points, but the argument is seen to be unaffected by that. It is also immaterial whether or not the duct axis is straight, provided that ρu is interpreted as the appropriate average, and the presence of solid duct walls is not essential either, provided only that they are impermeable and fixed in space. In general steady fluid motion, let C denote any smooth, closed, simple curve such that the velocity does not vanish, and is not tangent to C, at any point of C. A unique streamline can then be traced from each point of C and, by the smoothness convention, there is a neighborhood of C in which none of them encounters a stagnation point and in which the set of all points of such streamlines is a regular surface. Such a surface is called a *stream tube*, and on it $\mathbf{v} \cdot \mathbf{n} = 0$ at every point. With appropriate definitions of cross-section and average normal component of the mass-flux vector $\rho \mathbf{v}$, conservation of mass then implies (4.6) for the stream tube. In the incompressible case (4.7) follows, so that the average velocity is smaller where the cross section of the stream tube is larger.

Two-dimensional fluid motion is usually defined as motion such that a Cartesian coordinate system exists with respect to which ρ and $\mathbf{v}$ are independent of x_3, and $v_3 \equiv 0$; and this definition will also be adopted in the following. For such motion it is usual to write $x_1 = x$, $x_2 = y$ and $v_1 = u$, $v_2 = v$, and physical quantities are generally measured per unit "span"—meaning unit distance in the x_3-direction. If the motion is also steady, or incompressible, it can be represented in terms of a *stream function* ψ (see Problem 4.1) which measures mass flow rate and is such that the streamlines are identical with the lines $\psi = \text{const}$.

It is useful to remember from time to time that a two-dimensional motion does not exist in the real world; it is an approximation concept the plausibility of which is occasionally misleading. Similarly, steadiness is an asymptotic concept.

Axially symmetrical motion is defined as motion such that a cylindrical coordinate system x, r, φ exists with respect to which ρ and $\mathbf{v}$ are independent of the angular variable φ. The velocity component v_φ in the direction of increasing φ is called swirl velocity, and the meaning of axial symmetry is not very generally interpreted to imply $v_\varphi \equiv 0$.

Problem 4.1. Show from (4.2), without using (4.3), that a stream function $\psi(x, y)$ exists, in two-dimensional steady flow such that the difference between the respective values of ψ at any two points of a connected (see Section 5) fluid domain Ω_t equals the mass flow rate across an arbitrary (rectifiable) curve segment $\subset \Omega_t$ joining the two points. Deduce that $\mathbf{v} \cdot \text{grad } \psi = 0$ and [if $\rho \mathbf{v} \in C^1(\Omega_t)$] $\partial(\rho u)/\partial x + \partial(\rho v)/\partial y = 0$.

Problem 4.2. An open connected (see Section 5) set $\Omega \subset E^3$ is occupied by fluid in steady, axially symmetrical flow. Show from (4.2) without using (4.3) that a function $\psi(x, r)$ exists which is a stream function in a sense analogous to that defined in Problem 4.1. (The lines $\psi(x, r) = \text{const}$ are called *meridian streamlines*.) Deduce that $\mathbf{v} \cdot \text{grad } \psi = 0$ and (if $\rho \mathbf{v} \in C^1(\Omega)$)

$$(4.8) \qquad \partial(\rho v_x)/\partial x + \partial(\rho v_r)/\partial r + \rho v_r/r = 0$$

for $r \neq 0$, where v_x and v_r denote respectively the velocity components in the directions of increasing x and r.

5.   Circulation

A *path* is defined as a continuous, piecewise smooth, oriented curve, i.e., a curve representable in terms of a parameter σ by $\mathbf{x} = \mathbf{x}(\sigma)$ on $[0, 1]$ with piecewise continuous first derivative (continuous, that is, apart from a finite number of finite discontinuities). The orientation of a path P is defined by the sense of σ increasing, and the same curve traced in the opposite sense is called the path $-P$. A path is called *simple* if it does not cross itself, i.e., if $\sigma_2 \neq \sigma_1$ implies $\mathbf{x}(\sigma_2) \neq \mathbf{x}(\sigma_1)$. A set $D \subset E^3$ is *connected* means that, whenever $\mathbf{x}_1 \in D$ and $\mathbf{x}_2 \in D$, then there is a path $\subset D$ leading from $\mathbf{x}_1$ to $\mathbf{x}_2$. For instance, a spherical ball (i.e., the interior of a spherical surface) is connected, but the set consisting of two disjoint spherical balls is not. Similarly, the union of two spherical surfaces is connected if and only if the two surfaces have a common point.

A *circuit* is a path such that $\mathbf{x}(1) = \mathbf{x}(0)$, and is simple if $0 < \sigma_1 < \sigma_2 < 1$ implies $\mathbf{x}(\sigma_2) \neq \mathbf{x}(\sigma_1)$. The *circulation* of a circuit C is defined as

$$\Gamma(C) = \int_C \mathbf{v} \cdot d\mathbf{x} = \int_0^1 \mathbf{v} \cdot (d\mathbf{x}/d\sigma)\, d\sigma.$$

A circuit is *reducible* in a set D means that it is the whole boundary of an oriented surface $\subset D$. (A more precise definition is given in Appendix 5.) In practice, a sufficient test is often that the circuit can be deformed continuously, and entirely within D, until it has shrunk to a point of D. For instance, if D is the portion of the surface of the globe between the tropics of Cancer and Capricorn, then any sufficiently small circuit C in D is reducible in D (Fig. 5.1), but there are circuits (such as the equator) which are not.

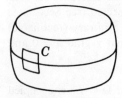

Fig. 5.1.

A set D is simply connected means that every circuit $\subset D$ is reducible in D. For example, the interior and exterior of a sphere are each simply connected, but those of a torus are not. Thus the fluid domain of Fig. 4.3 is simply connected, but that of Fig. 6.4 is not.

Fluid motion on a domain $\Omega_t \subset E^3$ is **irrotational** means that the circulation $\Gamma(C) = 0$ for every *reducible* circuit $C \subset \Omega_t$.

This imples—if x_1 and x_2 are any two points of a connected and simply connected domain Ω_t—that

$$\int_{x_1}^{x_2} \mathbf{v} \cdot d\mathbf{x}$$

is independent of the path of integration in Ω_t from x_1 to x_2 and hence is a function only of x_1, x_2, and t. But if x_0 is any third point of Ω_t,

$$\int_{x_1}^{x_2} \mathbf{v} \cdot d\mathbf{x} = \left(\int_{x_0}^{x_2} - \int_{x_0}^{x_1} \right) \mathbf{v} \cdot d\mathbf{x},$$

and therefore

(5.1) $$\int_{x_1}^{x_2} \mathbf{v} \cdot d\mathbf{x} = \phi(x_2, t) - \phi(x_1, t).$$

A *velocity potential* $\phi(\mathbf{x}, t)$ is thus defined (apart from an arbitrary additive constant) on any irrotational, simply connected fluid domain, and

(5.2) $$\mathbf{v} = \operatorname{grad} \phi.$$

(Some authors define this potential so that $v = -\operatorname{grad} \phi$, and this leads occasionally to a frustrating search for the source of a discrepancy in results based on information from different books.) Note that, if ϕ_1 and ϕ_2 are two potentials on the same domain, then $\phi_1 + \phi_2$ is also a potential on that domain; by (5.2), its velocity field is obtained from those of ϕ_1 and ϕ_2 by vector addition. If the motion is also incompressible, then from (4.5) and (5.2), ϕ satisfies Laplace's equation

(5.3) $$\nabla^2 \phi \equiv \operatorname{div} \operatorname{grad} \phi = \partial^2 \phi / \partial x_k \, \partial x_k = 0.$$

Some such motions are discussed in the next section.

It is of importance, particularly for aerodynamics, that many fluid domains are not simply connected. This does not affect the definition of irrotational motion, and, if it applies to the motion on a multiply connected fluid domain Ω_t (such as that of Fig. 6.4), then the velocity potential is defined, in the same manner as before, on any connected and simply connected subdomain. Thus (5.2) is valid everywhere on Ω_t. But (5.1) is valid only with qualification, for there are now irreducible circuits in Ω_t, and hence pairs x_1, x_2 of points for which the integral (5.1) may depend on the path of integration (Fig. 5.2). The velocity potential may therefore not be definable—whether by (5.1) or (5.2)—as a single-valued function on the whole domain Ω_t. However, the multivaluedness of ϕ then arising from the definition (5.1) is acceptable and, in fact, observable, as may be seen from the following example.

Consider steady, two-dimensional irrotational motion past an airfoil (i.e., a cylinder of infinite span, Fig. 5.2; of course, we mean the nearly two-dimensional motion which can plausibly exist near the center of a large wind tunnel spanned by the airfoil; the outer boundary of the fluid domain is not shown in the figure). It is plausible that the respective circulations of different,

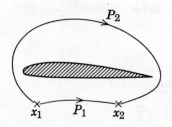

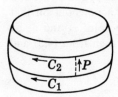

Fig. 5.2.

irreducible circuits in this fluid domain may be related. Generally, two circuits, C_1, C_2, in a set D are *homologous* (or reconcilable) means that there is a path $P \subset D$ such that $C_1 \cup P \cup (-C_2) \cup (-P)$ is a circuit reducible in D. In Fig. 5.1, e.g., any two circles of latitude in the part D of the surface of the globe between the tropics are homologous (if both are oriented east to west, say—a segment of circle of longitude can then serve as the path P, Fig. 5.3),

Fig. 5.3.

but no circle of latitude is homologous to a circuit reducible in D. It follows that the respective circulations of a pair of homologous circuits in a fluid domain always differ by the circulation of a reducible circuit. In irrotational motion, therefore, any two homologous circuits have the same circulation.

In the domain outside an airfoil (Fig. 5.2), whether two irreducible circuits are homologous depends only on (i) their orientations and (ii) how many times each goes around the airfoil. If this motion is irrotational, therefore, the circulation of *any* circuit must be an integer multiple of a single quantity Γ, which is called briefly the circulation of the airfoil. It will be seen in Section 13 that the lift L per unit span exerted by (ideal) fluid on the airfoil is $L = -\rho V \Gamma$, where V is the velocity far from the airfoil, so that $-V$ is the flight speed simulated in the tunnel. If the airfoil experiences a nonzero lift force, the potential is therefore infinitely many-valued, increasing by $\mp L/(\rho V)$ every time we go once around the airfoil.

More generally, a finite number of circuits C_i, the union of which is a connected set, may be considered to generate all the circuits obtained by tracing successively some, or all, of the C_i a number of times around in either sense. A connected set $D \subset E^3$ is *n-tuply connected* means that there are n basic circuits $C_i \subset D$ which are mutually nonhomologous in D and such that an arbitrary circuit $\subset D$ is homologous in D to a circuit generated by the C_i. (The fluid domains of Figs. 6.4 and 6.10 are thus seen to be doubly connected; so is the interior of a torus.) For irrotational motion on an n-tuply connected domain Ω_t there exist $n - 1$ functions $\Gamma_i(t)$, the circulations of the irreducible generating circuits, and the potential ϕ is single-valued on Ω_t at time t if and only if $\Gamma_i(t) = 0$ for every i.

Appendix 5

Homology in E^3 [5, 6]. An elementary 2-chain is the image, under a continuously differentiable map T into E^3, of a plane triangle the vertices of which are numbered 1, 2, and 3, say. (It is conventional to denote the map and the image by the same symbol.) For integration over a surface, its orientation (sense of the normal) may be relevant, and the numbering of the vertices defines an orientation for the triangle. If a map T' gives the same image as T, except that the numbers of just two of the three vertex-images are interchanged, one therefore writes $T' = -T$. Since a regular surface (Appendix 3) can be divided into a finite number of such triangle images, and since an integral over the surface is then the sum of integrals over the triangle images, it is natural to consider sums of elementary 2-chains directly. A general 2-chain is defined as a linear combination $c = \sum_{i=1}^{n} a_i T_i$ of such maps T_i of the triangle into E^3, with integer coefficients a_i.

An elementary 1-chain is the image, under a continuously differentiable map into E^3, of a straight line segment of which the ends are numbered; the orientation is from the lower to the higher number. A general 1-chain is, again, a linear combination of elementary 1-chains with integer coefficients. A path is, by its definition, seen to be a 1-chain. An elementary 0-chain, finally, is the image of a point under such a map. The boundary of an n-chain is therefore an $(n - 1)$-chain. More precisely, if T is again an elementary 2-chain, let $\partial_k T$ denote the elementary 1-chain which is the restriction of T to the triangle side opposite the vertex numbered k. Then the boundary of T is defined as the 1-chain $\sum_{k=1}^{3} (-1)^k \partial_k T = \partial T$. Similarly, for $n \geq 1$, the boundary of any n-chain $c = \sum a_i T_i$ is the $(n - 1)$-chain $\partial c = \sum a_i \sum (-1)^k \partial_k T_i$.

Two n-chains c_1, c_2 are homologous in a connected set $D \subset E^3$ means that there is an $(n + 1)$-chain $c \subset D$ the boundary of which is $\partial c = c_1 - c_2$. An n-chain is reducible in D means that it is the boundary of an $(n + 1)$-chain in D.

It may be mentioned in passing that a chain c such that $\partial c = 0$ is called a cycle. For instance, the boundary ∂T of the elementary 2-chain T has as its own boundary $\partial(\partial T)$ a linear combination of three 0-chains, each of which appears in the sum once with coefficient 1 and once with coefficient -1. More generally, a circuit is, by its definition, seen to be a 1-cycle. A standard example of a 2-cycle is a closed polyhedron. Some of the corollaries of Section 7 will be seen to be theorems on cycles.

6. Some Potential Flows

It has just been shown that, if a fluid motion be irrotational and incompressible, then the velocity field is the gradient of a potential ϕ satisfying Laplace's equation (5.3). Of course, it is a question of dynamics, to be discussed in the following chapters, under what circumstances a motion may actually be expected to be irrotational and incompressible. Meanwhile, some aspects of the preceding sections may be illustrated by a look at some solutions of Laplace's equation, interpreted as irrotational and incompressible motions, bearing in mind the kinematic boundary condition (3.4). Some solutions will look plausible as fluid motions, while others will look less plausible.

It may be noted that (5.2) and (5.3) do not involve the time. A time-dependent velocity potential ϕ may satisfy (5.3) at every t; or again, a motion may be irrotational and incompressible only at some times. It will be best, at this stage, to avoid such questions and to restrict the discussion to an interpretation of the potentials in terms of steady flows. To avoid lengthy calculations, moreover, it is best to start with two-dimensional flows, to which the elegant tools of complex analysis can be applied. (An elementary knowledge of that subject will suffice; and even in its absence, the non-analytical part of the discussion will be intelligible.) Following general usage the notation $x_1 = x, x_2 = y, v_1 = u, v_2 = v, (v_3 \equiv 0)$ and $z = x + iy$, $(i^2 = -1)$, is adopted for such flows.

In two-dimensional, incompressible flow a stream function $\psi(x, y)$ (Problem 4.1) exists, by (4.5), such that

(6.1) $$\partial\psi/\partial x = -v, \qquad \partial\psi/\partial y = u.$$

From (5.2)

$$\partial\phi/\partial x = u, \qquad \partial\phi/\partial y = v,$$

so that ϕ and ψ satisfy the Cauchy-Riemann equations [1]; hence

(6.2) $$\phi(x, y) + i\psi(x, y) = w(z)$$

is an analytic function [1] of $z = x + iy$. The velocity field is found from the *complex potential* $w(z)$ by noting that

(6.3) $$w'(z) \equiv dw/dz = \partial\phi/\partial x + i\,\partial\psi/\partial x = u - iv.$$

Since $\mathbf{v} \cdot \text{grad } \psi = 0$ (Problem 4.1), moreover, the boundary condition (3.4) is satisfied, with $\mathbf{V} = 0$, on any streamline $\psi(x, y) = \text{const}$; kinematically, therefore, any such curve can be interpreted as a solid boundary of the fluid.

The simplest example is $w = z$, i.e., $\phi = x, \psi = y, u = 1, v = 0$, which represents a uniform flow of unit velocity magnitude in the direction of increasing x.

Another example is

(6.4) $$w = z^2 = x^2 - y^2 + 2ixy.$$

Thus $\phi = x^2 - y^2$, $\psi = 2xy$, and $u = 2x$, $v = -2y$, by (6.3). The origin, $x = y = 0$, is a stagnation point ($u = v = 0$). Figure 6.1 shows the streamlines and the equipotential lines (dashed) in the upper half-plane ($y > 0$). Any hyperbola $xy = $ const can be interpreted as a solid boundary, and in particular, any coordinate half-axis may be so interpreted.

First choose the x-axis as solid boundary. Then the restriction of (6.4) to the upper half-plane ($y > 0$) has the interpretation of a steady stream impinging normally on a plane wall (Fig. 6.1).† Next choose the y-axis as

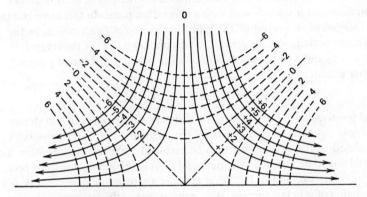

Fig. 6.1. Streamlines and equipotential lines (dashed) of the potential (6.4) for $y \geq 0$.

solid wall. The restriction of (6.4) to the right half-plane ($x > 0$; illustrated by Fig. 6.1 turned clockwise through 90° and with the streamline arrows reversed) then has the interpretation of a head-on meeting of two opposing streams along a wall and their coalescence into a stream leaving at right angles to the wall. It is not quite as plausible a physical picture as that of the impinging stream in the upper half-plane. Finally, choose the positive half-axes as solid walls. The restriction of (6.4) to the first quadrant ($x > 0, y > 0$; right half of Fig. 6.1) then has the interpretation of a flow past a concave, right-angled corner, entering in the direction of decreasing y and leaving in that of increasing x. (Just how plausible is it, physically? For further remarks see Section 27.)

More generally, consider a power series

(6.5)
$$w(z) = \sum_{n=2}^{\infty} a_n z^n$$

† The customary terminology reflects the dilemma between the three-dimensional, physical picture and the two-dimensional, analytical picture. Material, physical terms generally reflect the former, and geometrical terms, the latter. Thus the line $x = 0$ in the complex plane is called a wall, and the ignored x_3-axis of space is called a point.

with positive radius of convergence, and for ease of comparison assume $a_2 = 1$. Then, since the terms linear in z are absent, as $|z| \to 0$,

$$z^{-2}w(z) \to 1, \qquad z^{-1}w'(z) \to 2,$$

and sufficiently close to the origin, the velocity field of this more general potential is therefore approximated by the field of (6.4). The streamline $\psi = 0$ of (6.5) has four branches meeting at the stagnation point $z = 0$, each tangent to a coordinate half-axis. Suitable restrictions of (6.5) therefore represent flow past a smooth wall with a stagnation point. In this sense (6.5) will be recognized in several of the following examples (with $z - z_0$ in the place of z and with $a_2 \neq 1$, in general, which implies a shift of the stagnation point to $z = z_0$ and a counterclockwise rotation of Fig. 6.1 through $\frac{1}{2} \arg a_2$).

Another example is

(6.6) $$w(z) = U a (z/a + a/z)$$

with real positive constants U and a. It is a superposition of a uniform stream and a *dipole* potential characterized by the singularity of z^{-1} at $z = 0$. A neighborhood of the origin must thus be excluded from the fluid domain if Postulate IV is to be satisfied. The potential (6.6) is symmetrical with respect to the circle $|z| = a$, and in fact, that circle is a streamline $\psi = \text{const} = 0$. The restriction of w to the domain $|z| > a$ therefore has the interpretation of a flow past a circular cylinder of radius a. The streamline pattern is shown in Fig. 6.2. There are stagnation points $w'(z) = 0$ at (and only at) $z = \pm a$, and since $w''(\pm a) \neq 0$, series analogous to (6.5) converge in their respective neighborhoods, and considerations analogous to those following on (6.4) therefore apply to those neighborhoods. Figure 6.2 suggests that the velocity

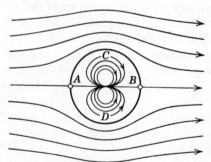

Fig. 6.2. Streamlines of the potential (6.6) of flow past a circular cylinder.

field is a plausible one upstream of the cylinder but is not equally plausible downstream. The qualms stem from the absence of a wake—the stream function of (6.6) is an even function of x.

The reader will have noticed that these examples all have unbounded fluid domains. This is a convenient limit for the description of velocity fields far from the outer boundary of the fluid. In the case of (6.6), e.g., the strict physical picture is that of a domain with an outer boundary component, $\partial\Omega_f$, and an inner one, $\partial\Omega_1 = \{|z| = a\}$, such that the minimum R of $|z|$ on $\partial\Omega_f$ is large compared with a. The mathematical picture is the limit of $w(z/a)$ as $R/a \to \infty$, so that the far boundary $\partial\Omega_f$ disappears. The boundary conditions at $\partial\Omega_f$ must then be represented by *conditions at infinity* which are their proper limit. This limit is sometimes obvious, but in other cases its correct formulation may require sophisticated physical and mathematical judgment, and this book will not attempt to come to grips with the problem in a general manner. In this section, in which only the interpretation of given analytic functions in terms of irrotational, incompressible flows is at issue, we may be content with noting possible interpretations of the limit of $w'(z)$ as $|z| \to \infty$.

Thus for (6.6), $\lim_{|z| \to \infty} w'(z) = U$, so that the velocity field, far from the cylinder, approximates a uniform stream. It thus comes arbitrarily close, at sufficiently large $|z|$, to satisfying, e.g., the kinematic boundary condition (3.4) on walls parallel to the x-axis. For (6.4), however, $|w'| \to \infty$ as $|z| \to \infty$, which admits no simple interpretation and thereby underlines that the significance of (6.4) lies in the approximation it furnishes to (6.5).

The potential

$$(6.7) \qquad\qquad w(z) = m \log z$$

with real constant m is called a *source* (and more particularly, a source if $m > 0$, but a sink if $m < 0$). With polar coordinates in the complex plane, $z = r \exp i\varphi$, $\phi = m \log r$, $\psi = m\varphi$, and by (6.3), the velocity is purely radial, of magnitude m/r, and directed away from (toward) the origin for a source (sink). To satisfy Postulate IV, the origin must again be excluded, and to make ψ single-valued (6.7) must be restricted to a definite branch of the logarithm. Since $\psi/\varphi = $ const, any two rays from the origin may be chosen as solid walls. The restriction of (6.7) to either of the sectors between those rays may then be interpreted as a flow between divergent, plane walls (with the branch cut of the logarithm chosen in the other sector). It is a limiting flow; the field close to $z = 0$ is not easily interpreted in terms of (3.4). For a sink, the fluid is sucked radially inward, and the limiting flow is a plausible one for any opening angle $\leq 2\pi$ of the sector. But for a source, the flow is radially outward and is plausible only for small opening angles; otherwise, experience suggests a jet. More direct interpretations are obtainable for linear combinations of sources and other potentials (Problems 6.1 and 6.2).

The potential

$$(6.8) \qquad\qquad w(z) = i \log z$$

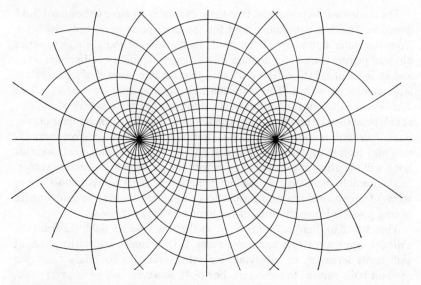

Fig. 6.3. Streamlines and equipotential lines of the potential flow of a source and sink of equal strength. (Observe that the equipotential lines are identical with the streamline of Fig. 8.1.)

is called a *vortex*. Its streamlines are the circles about the origin, and the fluid rotates clockwise about the origin, but with velocity magnitude $|z|^{-1}$ increasing inward, not like a solid body. To satisfy Postulate IV, the origin must again be excluded, but the stream function $\psi = \log |z|$ is single-valued and a branch cut is not needed. The domain $|z| > 0$ is doubly connected (Section 5), and any circuit around the origin has circulation -2π (a counterclockwise orientation of any circuit is assumed, unless the contrary is stated explicitly). Any circle $|z| = $ const > 0 may be chosen as solid boundary for a direct interpretation of (6.8), but the main significance of the vortex potential is an asymptotic one (Section 8).

Like (6.7), moreover, (6.8) may be fruitfully combined with other potentials. For instance,

$$(6.9) \qquad w(z) = Ua\left(\frac{z}{a} + \frac{a}{z}\right) + \frac{i\Gamma}{2\pi} \log \frac{z}{a}$$

with real constants U, a, Γ has the circle $|z| = a$ as streamline $\psi = 0$, like (6.6), and can thus be interpreted as flow, with circulation $-\Gamma$, about a circular cylinder immersed in a uniform stream (Fig. 6.4 and Problem 6.3). Joukowski's formula (Section 5) indicates an aerodynamic lift $L = \rho U\Gamma$ per unit span in the y-direction on the cylinder. But the velocity field is still

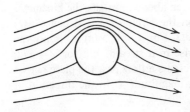

Fig. 6.4. Streamlines of a potential flow with circulation past a circular cylinder.

symmetrical with respect to the y-axis, so that there is no indication of a wake.

Many more examples can be given [7, 8], and a fruitful method for constructing them is the principle of *conformal mapping*, based on the chain rule for analytic functions. If $\zeta(z)$ is an analytic function on a domain Ω onto a domain D_ζ, and if $w_1(\zeta)$ is analytic on D_ζ, then $w(z) = w_1(\zeta(z))$ is analytic, and hence a complex potential, on Ω. Moreover, w_1 and w have the same circulation for corresponding circuits, and if im $w = $ const on a curve in the z-plane, then im $w_1 = $ const on the image of that curve under $\zeta(z)$; i.e., streamlines are mapped on streamlines. Thus if we wish to satisfy the kinematic condition (3.4) on the boundary $\partial\Omega$ of a fluid domain, it suffices to map Ω onto a domain D_ζ on the boundary of which (3.4) can be satisfied conveniently. The standard domains D_ζ used for that purpose are the upper half-plane and the unit disc (or the exterior of the unit circle); the Riemann mapping theorem assures the existence of suitable functions $\zeta(z)$ for an ample class of fluid domains Ω.

To illustrate the method, consider

$$(6.10) \qquad z(\zeta) = t + a^2/t, \qquad t = \zeta e^{i\alpha},$$

with real constants a and α on the exterior D_ζ of the circle $|\zeta| = |t| = a > 0$. It maps infinity on itself, since $z/\zeta \to e^{i\alpha}$ as $|\zeta| \to \infty$, and maps the circle $|\zeta| = a$ onto $z = 2a \cos \tau$, where $\tau = \arg t = \alpha + \arg \zeta$. Since $z(\zeta)$ is analytic on D_ζ, it therefore maps this domain onto the domain Ω, which is the z-plane cut along the real axis from $z = -2a$ to $z = 2a$. And since $dz/d\zeta$ does not vanish on $|\zeta| > a$, this map is $1:1$ and the inverse function $\zeta(z)$ is analytic on Ω. Hence, if $w_1(\zeta)$ is the potential of any flow past a circular cylinder of radius a, then $w(z) = w_1(\zeta(z))$ is the potential of a flow on Ω past a flat plate of chord $4a$. The complex velocity $u - iv$ on Ω is

$$(6.11) \qquad w'(z) = w_1'(\zeta) \, d\zeta/dz,$$

and in particular, $w'(z)/w_1'(\zeta) \to e^{-i\alpha}$ as $|z| \to \infty$, so that the velocities at infinity differ only in direction.

Begin with the potential (6.6),

$$w_1(\zeta) = u(\zeta + a^2/\zeta),$$

for flow without circulation on D_ζ. It was seen above to satisfy the kinematic boundary condition (3.4) (with $\mathbf{V} = 0$) on the circle ∂D_ζ, and hence $w(z)$ satisfies the same condition on the plate $\partial\Omega$. For $\alpha = 0$, $w(z)$ is simply the uniform flow past the plate placed edgewise in a stream. But for $\alpha = \pi/2$, $w'(z) \to -iU$ as $|z| \to \infty$, so that $w(z)$ represents a flow past the plate placed broadside (Fig. 6.5) in a stream of velocity magnitude U in the direction of

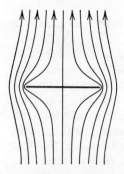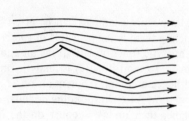

Fig. 6.5. Streamlines of a potential flow past a plate immersed broadside in a stream.

Fig. 6.6. Streamlines of a potential flow past a plate.

increasing y. And generally $w(z)$ represents a flow past the plate at *incidence* α, meaning that the velocity at infinity makes the angle α with the direction of the plate (Fig. 6.6, where the z-plane has been turned clockwise through the angle α). The complex velocity is

$$w'(z) = Ue^{-i\alpha}(t^2 - a^2e^{2i\alpha})/(t^2 - a^2),$$

so that $t = \pm ae^{i\alpha}$, $z = \pm 2a\cos\alpha$ marks the stagnation points (for $0 < \alpha \le \pi/2$), but for $\alpha > 0$ the velocity fails to exist at $t = \pm a$, $z = \pm 2a$, i.e., at the plate edges (Fig. 6.7). Indeed, since the plate is of zero thickness, the flow is clearly a limiting motion.

Its implausibility at one of the plate edges can be removed at small incidence α by the help of (6.9),

(6.12) $$w_1(\zeta) = U(\zeta + a^2/\zeta) + (i\Gamma/2\pi)\log\zeta,$$

which will give a flow with circulation $-\Gamma$ about the plate without altering the velocity at infinity. It then becomes possible to satisfy Postulate IV (for $0 \le \alpha < \pi/2$) at one of the plate edges, say at $z = +2a$. For this it is necessary, by (6.11), that $w_1'(\zeta) = 0$ at the point $\zeta = ae^{-i\alpha}$ corresponding to $z =$

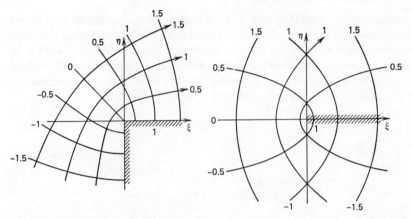

Fig. 6.7. Streamlines and equipotential lines of potential flows past a right-angled corner (left) and a plate edge (right).

$2a$, where $dz/d\zeta = 0$. From Problem 6.3, or directly from (6.12), this requires $\Gamma = 4\pi Ua \sin \alpha$. That is also sufficient, because (6.10) to (6.12) then give

$$w'(z) = Ue^{-i\alpha}(t^2 - a^2)^{-1}(t^2 + 2iate^{i\alpha}\sin \alpha - a^2e^{2i\alpha})$$
$$= Ue^{-i\alpha}(t + a)^{-1}(t + ae^{2i\alpha}),$$

which exists at $t = a, z = 2a$. The lift per unit span is then $L = -\rho U\Gamma = 4\pi\rho U^2 a \sin \alpha$, by Joukowski's theorem (Section 11). The difficulty at the other plate edge is not removed (Fig. 6.8, where the z-plane has again been

Fig. 6.8. Streamlines of a potential flow with circulation past a plate.

turned clockwise through the angle α), so that $w(z)$ must still be interpreted as a limiting motion. It is a useful one in the sense that the lift is an approximation to the real lift on a plate at small incidence α.

A potential for which the velocity field is everywhere plausible can be obtained for small incidence by using a potential analogous to (6.12) representing flow past a circular cylinder C enclosing the cylinder $|\zeta| = a$ [7]. If the circles C and $|\zeta| = a$ touch at $\zeta = ae^{-i\alpha}$, and if $(|\zeta| - a)/a$ is small on C, then (6.10) maps C on an Ω-contour with cusped "trailing" edge resembling an airfoil (Fig. 6.9). The difficulty at the nose of the contour is now avoided, since $w(z)$ is analytic in its neighborhood. The difficulty at the sharp trailing edge (Fig. 6.9) can again be avoided by insisting on Postulate IV even there. As before, this turns out to be possible for just one value of the circulation, whence the lift is also determined by Joukowski's theorem (Section 11). The corresponding streamline pattern is shown in Fig. 6.10.

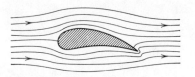

Fig. 6.9. Streamlines of a potential flow without circulation past an airfoil.

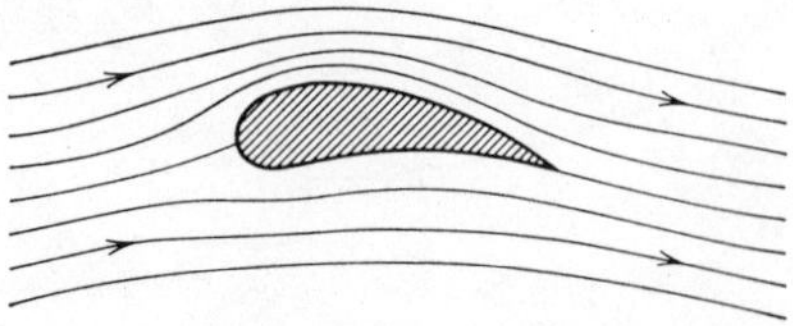

Fig. 6.10. Streamlines of the potential flow past an airfoil for which the Kutta-Joukowski condition is satisfied.

The *Kutta-Joukowski condition* that Postulate IV apply at a sharp trailing edge is equally useful for many other flows. Nonetheless it is a rule of limited scope, as becomes clear when the incidence is increased further and further (Fig. 6.11). Intuition indicates that a stage must come where the flow is more

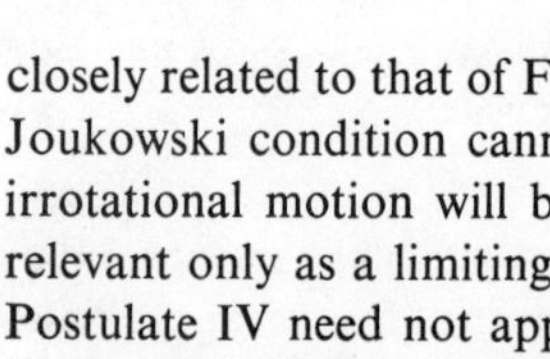

Fig. 6.11.   Airfoil at high incidence.

closely related to that of Fig. 6.12 than those of Figs. 6.6 and 6.8. The Kutta-Joukowski condition cannot, in fact, be justified by Postulate IV, because irrotational motion will be seen in the following chapters to be physically relevant only as a limiting motion corresponding to zero viscosity, to which Postulate IV need not apply uniformly. The real justification of the Kutta-Joukowski condition rests on boundary layer theory (Chapter 4).

It is instructive to note that the method of conformal mapping yields even flows of the type shown in Figs. 6.12 and 4.1, if assumptions are added to make the potential determinate [7–9].

Three-dimensional potential flow presents greater analytical difficulty, since complex analysis is not directly applicable, but some of the examples

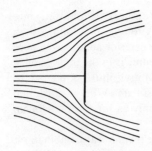

Fig. 6.12. Streamlines of a two-dimensional potential flow past a plate (compare with Fig. 6.5).

just discussed are readily extended to axially symmetrical flow. It is then convenient to use cylindrical coordinates x (for distance along the axis of symmetry) and R (for distance from this axis), and to denote the velocity components in the directions of increasing x and R, respectively, by v_x and v_R. (Only flows without swirl will be discussed.) From (5.2),

$$v_x = \partial\phi/\partial x, \qquad v_R = \partial\phi/\partial R.$$

Since the flow is axially symmetrical and incompressible, there is also a (meridian) stream function $\psi(x, R)$ (Problem 4.2) such that

$$v_x = R^{-1}\,\partial\psi/\partial R, \qquad v_R = -R^{-1}\,\partial\psi/\partial x.$$

A uniform flow must be axial, on account of the assumed symmetry, and its velocity potential is again $\phi = Ux$, if U is the velocity in the direction of increasing x; its stream function is $\psi = UR^2/2$. The analog of the stagnation-point flow (6.4) is

$$(6.13) \qquad \phi = R^2/2 - x^2, \qquad \psi = -R^2 x,$$

with velocity components $v_x = -2x$ and $v_R = R$. Its restriction to $x < 0$ (or $x > 0$) has the interpretation of an axial stream impinging on the plane wall $x = 0$ (on which (3.4) is satisfied), which converts the stream into a radial flow. As in the two-dimensional case, the velocity magnitude grows beyond bound as $x^2 + R^2 \to \infty$, so that the primary significance of (6.13) must again be that it represents the first terms of a Taylor series with respect to x and R in the neighborhood of stagnation points of flows with more direct interpretation.

The potential

$$\phi = -mr^{-1}, \qquad r^2 = x^2 + R^2, \qquad m = \text{const}$$

is called a (three-dimensional) *source* at the origin (especially if $m > 0$, and often a sink if $m < 0$). Since this potential is spherically symmetrical, the velocity is entirely radial; its magnitude is $|m|r^{-2}$. The volume flow rate across any sphere $r = \text{const} > 0$ is therefore $4\pi m$ (outward). Moreover, since the flow is incompressible and satisfies Postulate IV for $r > 0$, it follows that, if S is any closed surface separating space into the exterior and interior of S, then the volume flow rate across S is $4\pi m$ if the source is in the interior of S, and zero if it is in the exterior. The source ($r = 0$) itself must, of course, be excluded from the fluid domain, in order that Postulate IV be satisfied, and altogether, the same asymptotic qualifications apply as for the two-dimensional source discussed earlier.

More direct interpretations are again obtainable from linear combinations of sources and other potentials (e.g., Problem 6.4), and for this it is helpful to employ also spherical coordinates r and θ (Fig. 6.13). If the velocity components in the directions of increasing r and θ are denoted respectively by v_r and v_θ, the connection between the two representations is given by (Fig. 6.13)

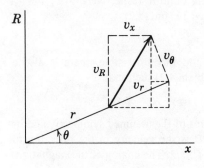

Fig. 6.13. Velocity components with respect to cylindrical and spherical coordinates.

$$x = r \cos \theta, \qquad R = r \sin \theta,$$
$$v_x = \partial\phi/\partial x = R^{-1} \, \partial\psi/\partial R = v_r \cos \theta - v_\theta \sin \theta,$$
(6.14) $\qquad v_R = \partial\phi/\partial R = -R^{-1} \, \partial\psi/\partial x = v_r \sin \theta + v_\theta \cos \theta,$
$$v_r = \partial\phi/\partial r = (Rr)^{-1} \, \partial\psi/\partial\theta = v_x \cos \theta + v_R \sin \theta,$$
$$v_\theta = r^{-1} \, \partial\phi/\partial\theta = -R^{-1} \, \partial\psi/\partial r = -v_x \sin \theta + v_R \cos \theta.$$

The stream function for the source is thus seen to be

$$\psi = -m \cos \theta = -mx(R^2 + x^2)^{-1/2}.$$

The potential

$$\phi = \mu \, \partial(r^{-1})/\partial x = -\mu x/r^3, \qquad \mu = \text{const}$$

is called a *dipole* at the origin directed in the sense of increasing x (if $\mu > 0$). The corresponding stream function is

$$\psi = \mu \, \partial(\cos \theta)/\partial x = \mu R^2/r^3.$$

In analogy to (6.6), we may combine this with a uniform flow to get

(6.15) $\phi = Ux\,(1 - \mu r^{-3})$, $\psi = \tfrac{1}{2}UR^2(1 + 2\mu U^{-1}r^{-3})$,

and if $\mu < 0$ then $\psi = 0$ on $r = (-2\mu/U)^{\frac{1}{3}} = a$, i.e. (3.4) is satisfied on $r = a$, and the restriction of (6.15) to $r > a$ therefore has the interpretation of potential flow past a sphere of radius a, with uniform velocity at infinity (Fig. 6.14). The potential is odd in x, so there is still no indication of a wake.

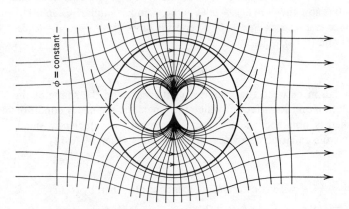

Fig. 6.14. Meridian streamlines and equipotential lines of an axially symmetrical, irrotational, incompressible flow past a sphere.

A more general representation of axially symmetrical, incompressible potential flow past bodies of revolution is obtainable from a continuous superposition of sources, i.e.,

$$\phi = Ux - \int_0^a r^{-1}m(s)\,ds, \qquad r^2 = R^2 + (x - s)^2$$

$$\psi = \tfrac{1}{2}UR^2 - \int_0^a r^{-1}(x - s)m(s)\,ds.$$

This corresponds to a *line source* on the segment $(0, a)$ of the axis producing a volume flow rate $4\pi m(x)$ per unit length. We may guess that such a potential will be interpretable as a flow past a finite body if $\int_0^a m(x)\,dx = 0$. To obtain convenient expressions for the velocity, assume that

(6.16) $m(0) = m(a) = 0$, $m(x) \in C^1[0, a]$

and integrate by parts to get

$$\phi(x, R) - Ux = \int_0^a m'(s)\,\sinh^{-1}\frac{s - x}{R}\,ds$$

for $R > 0$, whence by direct differentiation,

$$v_x - U = -\int_0^a r^{-1}m'(s)\,ds,$$

(6.17)

$$v_R = R^{-1}\int_0^a r^{-1}(x - s)m'(s)\,ds.$$

Observe that Rv_R tends to a limit as $R/x \to 0$ at the axis. Indeed, for $0 < s < x$, $r \to x - s$ if $R/(x - s) \to 0$, and $r/x \to 0$ if $R/x \to 0$ but $R/(x - s) \nrightarrow 0$, so that $m'(s) \to m'(x - r)$; and, similarly, $m'(s) \to m'(x + r)$ for $x < s < a$ as $R/x \to 0$. Therefore, since $(x - s)/r = -\partial r/\partial s$, and by (6.16) and (6.17),

$$Rv_R \to 2m(x) \quad \text{as} \quad R/x \to 0.$$

A simple relation between the source strength $m(x)$ and the shape of the body of revolution on the surface of which the kinematic boundary condition (3.4) is satisfied is obtained when the source strength is sufficiently small. If $R = R_s(x)$ is the equation of a streamline $\psi = $ const, then by (6.14)

$$dR_s/dx = R_s'(x) = v_R/v_x,$$

and if $|m'(x)| \le kU\varepsilon^2$ with constant k independent of ε, then by (6.17)

$$\left|\frac{v_x - U}{U}\right| \le k\varepsilon^2\left(\sinh^{-1}\frac{a - x}{R} + \sinh^{-1}\frac{x}{R}\right) = \varepsilon^2\,\delta(x, R).$$

Hence, on $R = R_s(x)$ for $0 < x < a$,

$$2\pi R_s R_s' \to 4\pi m(x)/U \quad \text{as} \quad \varepsilon^2\,\delta(x, R_s(x)) \to 0,$$

and if $R_s(x) = 0$ for $x \le 0$ and $x \ge a$, and $R_s(x) > \alpha\varepsilon x(a - x)$ on $(0, a)$, then $\varepsilon^2\,\delta(x, R_s) \to 0$ with ε, except at $x = 0$ and $x = a$. For sufficiently small thickness ratio $\varepsilon = (2/a)\max R_s(x)$, therefore, the field (6.17) of the line source potential can be interpreted as an approximation to a potential flow past a pointed, slender body of revolution of cross-sectional area $S(x) = \pi R_s^2$ and length a if

$$m(x) = US'(x)/2$$

and $\alpha\varepsilon x(a - x) < R_s(x) < \beta\varepsilon x(a - x)$, $|R_s'(x)| < \beta\varepsilon$, and $|R_s''(x)| < \beta\varepsilon$ with constants α and β independent of ε [10]. The approximation is nonuniform, however; it fails in very small neighborhoods of the nose and the tail.

Problem 6.1. Show that $w(z) = m \,\text{kog}\, z + Uz$, with real positive constants m and U and branch cut along the positive real axis, can be interpreted as a potential flow past a blunt-nosed body of infinite length and finite thickness. Find the asymptotic thickness, and the position of the nose, of the body.

Problem 6.2.   Show that

$$w(z) = Uz + m \log [(z + a)/(z - a)],$$

with real positive constants m, U, a and branch cut along the real axis from $z = -a$ to $z = a$, can be interpreted as a potential flow past a finite body with smooth contour. Find the thickness of the body and show that the thickness ratio (total thickness over total length) τ is such that $Ua\tau/m \to \pi$ as $m/(Ua) \to 0$. Formulate the limit in which this flow tends to that past a circular cylinder. (A convenient approach is to establish, by the help of the signature of the y-component of velocity, that a closed streamline passes through the stagnation points.)

Problem 6.3.   Find the stagnation points of (6.9) and show that (6.5) applies near them, unless $|\Gamma| = 4\pi Ua$. (It is best to use polar coordinates, $z = r \exp i\varphi$.)

Problem 6.4.   Show that the sum of the stream functions of a source of volume flow rate $4\pi m$ and of a uniform stream of velocity U may be interpreted as a stream function of a potential flow past a blunt-nosed, smooth, axisymmetric half-body of asymptotic thickness $2(m/U)^{\frac{1}{2}}$. Confirm that the first approximation to the potential, close to the nose, agrees with (6.13).

Appendix 6

Uniqueness of Potentials and Maximum Principle. The collection of examples of velocity potentials in Section 6 raises natural questions concerning uniqueness, and some simple, partial answers and related facts follow.

Theorem 6.1   There cannot be more than one irrotational motion of incompressible fluid (with potential $\phi \in C^1(\overline{\Omega})$ and $\in C^2(\Omega)$) on an open bounded set $\Omega \subset E^3$ with regular boundary surface $\partial\Omega$, for which $\mathbf{n} \cdot \mathbf{v}$ is prescribed on all of $\partial\Omega$ and for which the circulations of all irreducible, basic generating circuits (Section 5) are prescribed.

Theorem 6.2.   There cannot be more than one irrotational motion of incompressible fluid (with potential $\phi \in C^1(\overline{\Omega})$ and $\in C^2(\Omega)$) on an open bounded set $\Omega \in E^3$ with regular boundary surface $\partial\Omega$, for which $\mathbf{n} \wedge \mathbf{v}$ is prescribed on all of $\partial\Omega$ and ϕ is prescribed at one point of each connected component of $\partial\Omega$.

Both theorems can be established by the Energy Method of the theory of partial differential equations. Indeed, the kinetic energy of the fluid in Ω is

$$\tfrac{1}{2}\rho \int_\Omega |\mathbf{v}|^2 \, dV = \tfrac{1}{2}\rho \int_\Omega |\text{grad } \phi|^2 \, dV.$$

Since $\phi \in C^2(\Omega)$, $\text{div} (\phi \text{ grad } \phi) = \phi \nabla^2\phi + |\text{grad } \phi|^2$, and, by (5.3), $\nabla^2\phi \equiv 0$ on Ω. If ϕ is single-valued on Ω, then since $\phi \text{ grad } \phi \in C(\overline{\Omega})$, the Divergence theorem can be applied to obtain

$$(6.18) \qquad \int_\Omega |\mathbf{v}|^2 \, dV = \int_{\partial\Omega} \phi \mathbf{n} \cdot \text{grad } \phi \, dS.$$

For Theorem 6.1 let ϕ_1 and ϕ_2 be two potentials defined on Ω at the same time and then satisfying the hypotheses with the same boundary values and circulations. Then $\phi = \phi_2 - \phi_1$ also satisfies the hypotheses and (5.3), with $\mathbf{n} \cdot \operatorname{grad} \phi \equiv 0$ on $\partial\Omega$ and with zero circulation for all the basic, generating circuits. It follows (Section 5) that ϕ is single-valued on Ω, and hence, from (6.18), that the difference between the velocity fields grad ϕ_2 and grad ϕ_1 vanishes identically on $\overline{\Omega}$.

For Theorem 6.2 let ϕ again denote the difference between two potentials satisfying the hypotheses with the same boundary values. Then $\mathbf{n} \wedge \operatorname{grad} \phi \equiv 0$ on $\partial\Omega$, and since $\phi \in C^1(\overline{\Omega})$, ϕ must be identically constant on each connected component of $\partial\Omega$. Since $\phi_2 = \phi_1$ at a point of each such component, moreover, all these constants are zero. If ϕ is single-valued on Ω, (6.18) will therefore again establish the theorem. Now let C denote any one of the basic, generating circuits; since it is a cycle (Appendix 5), there is a 2-chain $S \subset E^3$ of which it is the boundary [6], and $S \cap \overline{\Omega}$ is a 2-chain of which the boundary is $C \cup C'$, where C' is a cycle in $\partial\Omega$. Since $\phi \equiv 0$ on $\partial\Omega$ and since C and $-C'$ are homologous, it follows that C has zero circulation.

The reader will have noticed that neither theorem refers to a boundary value problem of completely conventional type. It should also be mentioned that the values of $\mathbf{n} \cdot \mathbf{v}$ on $\partial\Omega$ in Theorem 6.1 cannot be prescribed quite arbitrarily; consideration of mass conservation, or of (4.3) and the divergence theorem, will show that the potential cannot exist unless these boundary values satisfy $\int_{\partial\Omega} \mathbf{n} \cdot \mathbf{v} = 0$.

It is even more relevant to fluid dynamics that the smoothness conditions on ϕ of Theorems 6.1 and 6.2 cannot be weakened appreciably. Those stated are, of course, amply covered by the smoothness convention. For limiting motions, however, they need not be satisfied, and the remaining conditions listed in Theorems 6.1 and 6.2 may then fail utterly to assure uniqueness!

These theorems can also be approached from the maximum principle of potential theory, which follows directly from

Theorem 6.3. (Mean Value Theorem of Potential Theory). Let S denote any sphere such that the ball interior to it is a subset of an open set $\Omega \subset E^3$. Then if $\phi \in C^1(\Omega)$ and is single-valued, and if $\nabla^2 \phi = 0$ on Ω, the mean of ϕ over S equals the value of ϕ at the center of the sphere.

Corollary 6.1 (Maximum Principle of Potential Theory). A maximum or minimum of ϕ cannot occur in Ω.

For the proof of Theorem 6.3 let r denote distance from the center of the sphere. The mean of ϕ over S is then

$$M(r) = \frac{1}{4\pi r^2} \int_S \phi \, dS = \frac{1}{4\pi} \int_0^{4\pi} \phi(r, \omega) \, d\omega,$$

where ω denotes the solid angle subtended at the center of the sphere. Since $\phi \in C^1(\Omega)$,

$$\frac{dM}{dr} = \frac{1}{4\pi} \int_0^{4\pi} \frac{\partial \phi}{\partial r} \, d\omega = \frac{1}{4\pi r^2} \int_S \mathbf{n} \cdot \operatorname{grad} \phi \, dS = 0$$

by the Divergence theorem.

Corollary 6.2. In nonuniform, irrotational motion of incompressible fluid, a maximum of $|\mathbf{v}|$ cannot occur in the interior of the fluid domain Ω if $\mathbf{v} \in C^2(\Omega)$.

For proof, let P denote any interior point of Ω and choose Cartesian axes so that the velocity has components $u, 0, 0$ at P. Since $\nabla^2\phi = 0$, also

$$\frac{\partial}{\partial x_1}\nabla^2\phi = \nabla^2\frac{\partial\phi}{\partial x_1} = 0$$

on Ω, and by Theorem 6.3, every neighborhood of P must contain another point at which $\partial\phi/\partial x_1 \geq u$ and therefore $|\mathbf{v}|^2 = \sum_1^3 (\partial\phi/\partial x_i)^2 \geq (\partial\phi/\partial x_1)^2 \geq u^2$.

It follows further [1, Vol. 2, p. 97] that, if the fluid domain is bounded and also $\mathbf{v} \in C(\overline{\Omega})$, then the maximum of $|\mathbf{v}|$ must occur on the boundary $\partial\Omega$ of the fluid domain.

Lemma 12.1 will imply further:

Corollary 6.3. In steady, irrotational flow of ideal fluid, a minimum of the (hydrodynamic) pressure cannot occur at an interior point of the fluid domain Ω if $\mathbf{v} \in C^2(\Omega)$.

Problem 6.5. Give an example of a simply-connected, bounded domain Ω and a nontrivial potential $\phi \in C^2(\overline{\Omega})$ for which $\nabla^2\phi \equiv 0$ on Ω and $\mathbf{n} \wedge \operatorname{grad} \phi \equiv 0$ on $\partial\Omega$.

Problem 6.6. Deduce Theorem 6.2 from the Maximum Principle, rather than from (6.18).

7. Vorticity

The tensor of velocity derivatives $\partial v_i/\partial x_k$ $(i, k = 1, 2, 3)$ may be split into its symmetrical and antisymmetrical parts, $e_{ik} = \frac{1}{2}(\partial v_i/\partial x_k + \partial v_k/\partial x_i) = e_{ki}$ and $\alpha_{ik} = \frac{1}{2}(\partial v_i/\partial x_k - \partial v_k/\partial x_i) = -\alpha_{ki}$. Being antisymmetrical, the tensor α_{ik} is related to a vector curl $\mathbf{v} = \boldsymbol{\omega}$, called *vorticity*, by

$$(7.1) \qquad \omega_i = \varepsilon_{ijk}\alpha_{kj} \qquad (i = 1, 2, 3),$$

where ε_{ijk} denotes the alternating tensor (i.e., $\varepsilon_{ijk} = \varepsilon_{jki} = -\varepsilon_{kji}$ and $\varepsilon_{123} = 1$). In terms of the velocity components $v_1 = u$, $v_2 = v$, and $v_3 = w$ with respect to Cartesian coordinates $x_1 = x$, $x_2 = y$, and $x_3 = z$,

$$(7.2) \qquad \omega_1 = \frac{\partial w}{\partial y} - \frac{\partial v}{\partial z}, \qquad \omega_2 = \frac{\partial u}{\partial z} - \frac{\partial w}{\partial x}, \qquad \omega_3 = \frac{\partial v}{\partial x} - \frac{\partial u}{\partial y}.$$

The immediate physical significance of the vorticity in the local kinematics of a fluid is best appreciated from Problem 7.1 below. It shows the vorticity to be a generalization appropriate to continuum kinematics of the angular velocity concept of rigid body kinematics. Being related to an antisymmetrical tensor and to rotation, the vorticity is a pseudo-vector; i.e., it transforms as a vector in all coordinate rotations except mirror transformations, where its transformation rules differ in sign from those of a vector. Rather than take repeated account of this distinction, we adopt the convention that only right-handed coordinate systems are to be employed in this book.

The most striking properties of vorticity are properties *in the large*, which flow from the remarkable theorems of Stokes and Kelvin to be discussed now and in Section 14, respectively. In Stokes' theorem and its corollaries, time-dependence plays no direct role, and in the rest of this chapter any fluid domain will usually be considered only at an (arbitrary) fixed time. On the other hand, a small extension of some notions outlined in Appendix 1 is needed. The reader will have noticed that these do not require the context of three-dimensional Euclidean space. In the plane, for instance, the concept of neighborhood of a point may be based on a circular disc centered at the point, rather than on a small spherical ball, to define open set, boundary, etc., literally as in Appendix 1. The disc neighborhood concept may next be extended to surfaces, to define open sets, etc., on them. A manifold is a set every point of which has such a disc neighborhood which also belongs to the set. A manifold-with-boundary is a set each point of which has either a disc neighborhood, or a half-disc neighborhood, which also belongs to the set. These are the concepts which will be referred to briefly as *surface* and its *boundary*. The conventional term *closed surface* will be used for a manifold which is a bounded set and the boundary of which is empty. (For definitions of other terms used below, see Sections 3, 5, and their appendices.)

Stokes' theorem. If S is a regular surface the boundary ∂S of which is a single circuit, and if $\mathbf{f}(\mathbf{x}) \in C^1(G)$, where G is open in E^3 and $G \supset S$, then

$$\int_S \mathbf{n} \cdot \operatorname{curl} \mathbf{f} \, dS = \int_{\partial S} \mathbf{f} \cdot d\mathbf{x},$$

where $\mathbf{n}$ is the unit normal on S related to the orientation of the circuit ∂S by the right-hand screw rule.

A proof will be found in many books on vector analysis and potential theory [e.g., 3, 5]. With $\mathbf{f} = \mathbf{v}$, the theorem relates vorticity and circulation, or more precisely, $\Gamma(\partial S)$ and the "vortex flux" $\int_S \boldsymbol{\omega} \cdot \mathbf{n} \, dS$ through S, and implies immediately

Corollary 7.1. If S is a closed regular surface in a fluid domain, then $\int_S \boldsymbol{\omega} \cdot \mathbf{n} \, dS = 0$.

It follows by the Divergence theorem that $\int_R \operatorname{div} \boldsymbol{\omega} \, dV = 0$ whenever $R \subset \Omega_t$ is the interior of a closed regular surface, and hence $\operatorname{div} \boldsymbol{\omega} = 0$ on any fluid domain on which $\operatorname{div} \boldsymbol{\omega}$ is continuous. Conversely, it is easy to verify that $\operatorname{div} \operatorname{curl}$ is a null-operator, and much of Corollary 7.1 can then be deduced by the Divergence theorem.

Corollary 7.2. The vortex flux $\int_S \boldsymbol{\omega} \cdot \mathbf{n} \, dS$ has the same value for all regular surfaces $S \subset \Omega_t$ which have the same single circuit as boundary, provided that the normals are consistently oriented.

This obvious corollary extends the theorem to surfaces which are not regular but may be adequately approximated by regular surfaces.

Corollary 7.3. Stokes' theorem applies also to any regular surface the boundary of which is the union of a finite number of circuits, provided they are all consistently oriented.

This follows immediately from the idea indicated in Fig. 7.1 in the case of any one practical application. A general proof follows from the same construction (Fig. 7.1), if the existence of a suitable system of auxiliary boundary

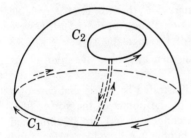

Fig. 7.1.

segments connecting all the circuits can be established. That is guaranteed for a surface which is triangulable in the sense of Appendix 5 (i.e., for a 2-chain). The corollary therefore holds for any such surface, not only for regular ones; the vortex flux $\int \boldsymbol{\omega} \cdot \mathbf{n} \, dS$ is then understood in a generalized sense defined by the circulation. Corollaries 7.1 and 7.2 are similarly valid for triangulable surfaces, even on the closure of the fluid domain.

Vortex line means trajectory of the field $\boldsymbol{\omega}$, i.e., a curve $\mathbf{x}(\sigma) \subset \Omega_t$ such that $d\mathbf{x}/d\sigma$ is a vector parallel to $\boldsymbol{\omega}(\mathbf{x}(\sigma))$ for every σ in the interval of definition of $\mathbf{x}(\sigma)$. By the smoothness convention (Section 3) and the fundamental theorem of ordinary differential equations [2], a unique vortex line passes through any point at which $\boldsymbol{\omega} \neq 0$. A vortex line cannot, therefore, end at any such point. However, the definition cannot be applied at a point where $\boldsymbol{\omega} = 0$, and since sets of such points are not uncommon it is important to be aware of much stronger results implied by Stokes' theorem.

If Δ is an open subset of Ω_t in which $\boldsymbol{\omega}$ does not vanish, then the union of the points of all the vortex lines intersecting Δ is called a *vortex bundle*. This definition, local to begin with, implies that a vortex bundle B is an open subset of Ω_t. Its boundary ∂B may intersect $\partial \Omega_t$, and the complement $\partial B' = \partial B - (\partial B \cap \partial \Omega_t)$ of that intersection with respect to ∂B is called a *vortex tube*. Such a tube cannot be crossed by a vortex line, for if $\boldsymbol{\omega} \neq 0$ at a point

$P \in \partial B'$, then a vortex line V through P is defined, and if V did intersect the bundle B, then $V \subset B$ by the definition. Hence there is no vortex flux through any subset of a vortex tube, and Corollary 7.3 implies

Corollary 7.4. If two circuits are subsets of a vortex tube and homologous therein, then they have the same circulation.

Another formulation is

Corollary 7.5. If a circuit in a vortex tube has nonzero circulation, then the circuit is irreducible in the vortex tube.

Indeed, if the vortex tube $\partial B'$ contains a circuit C, then $\partial B'$ is a triangulable surface by the smoothness convention (Section 3 and Appendix 5). If C were reducible in $\partial B'$, it would be the boundary of a triangulable surface $S \subset \partial B'$, whence the vortex flux $\int_S \boldsymbol{\omega} \cdot \mathbf{n} \, dS = 0$. But then by Stokes' theorem, C has zero circulation.

The thrust of these corollaries may be illustrated by the special case of a *simple vortex bundle*, meaning the union of the points of the vortex lines intersecting an open set $\Delta \subset \Omega_t$ which is connected and simply connected (Section 5). The vortex tube T of such a bundle is a hose-like surface, and Corollary 7.5 shows that there is a class of circuits $C_\alpha \subset T$ all of which have the same circulation and might be called "belts" of the bundle and tube. Their circulation Γ is a property of the bundle, called its strength; it is also the total vortex flux through any triangulable surface S such that $\partial S = C_\alpha$ for some α, by Corollary 7.2. The existence of such an invariant Γ indicates a quality of geometrical permanence in the bundle. Indeed, Corollary 7.5 states that a *simple vortex bundle of nonzero strength cannot end in the fluid domain*, because a belt cannot be the whole boundary of a subset of the bundle-boundary disjoint from $\partial \Omega_t$.

A vortex line, in turn, can be regarded (by the smoothness convention) as a limit of a sequence of simple vortex bundles, and when it is so defined, then it follows also that *a vortex line cannot end in a fluid domain* Ω_t. It must be either a closed curve or a curve connecting two points of $\partial \Omega_t$.

Problem 7.1. Show that if a body of fluid rotates as a solid body with angular velocity $\boldsymbol{\alpha}$, then the vorticity $\boldsymbol{\omega} \equiv 2\boldsymbol{\alpha}$ on the fluid domain.

8. Line Vortex

The theorems of the last section make possible a better attempt than in Section 6 to communicate a degree of "feel" for some aspects of real fluid motion. To this end, it is best to discuss vortex bundles embedded in irrotational motion. A velocity potential ϕ (Section 5) then exists on every open subset of the fluid domain on which the motion is irrotational, and it may be multivalued, even if the fluid domain (in contrast to the irrotational subdomain) is simply connected.

For convenience the discussion will refer to unbounded fluid domains, to be interpreted in accordance with the remarks on such domains in Section 6; in the following examples $|\mathbf{v}| \to 0$ as $|\mathbf{x}| \to \infty$. To begin with, the velocity fields discussed will be thought of as snapshots of motions, as in Section 7. Later the snapshots will be strung together into time sequences.

As a first example, consider a straight vortex bundle B of strength $\Gamma \neq 0$ in the fluid domain $\Omega_t = E^3$. Cylindrical coordinates x, R, and θ will be used such that $B \subset \{R < d\}$. On the complement of B, $\boldsymbol{\omega} \equiv 0$. The discussion can be greatly simplified by inquiring about the velocity field only at distances R such that $R/d \to \infty$. For the description of this "far field" it is best to choose a length l such that $d/l \to 0$ as the unit in terms of which x and R are measured. The motion is then irrotational, except on the axis $R = 0$ (meaning, strictly, $R/l = 0$) which has circulation Γ. The potential (6.8) describes such a motion;† in the cylindrical coordinates it is

$$(8.1) \qquad \phi = (2\pi)^{-1}\Gamma\theta.$$

As noted in Section 6, its streamlines are circles about the x-axis in planes normal to that axis, and the velocity magnitude is

$$(8.2) \qquad |\mathbf{v}| = |\Gamma|/(2\pi R).$$

It should be stressed that this represents a quite different view of the potential (6.8) than in Section 6. The axis $R = 0$ was excluded there from the fluid domain by Postulate IV. The present view does not exclude the axis from the fluid domain, but rather interprets (8.1) as an asymptotic representation of a straight vortex bundle seen from afar. It is worth observing that this uncertainty (or flexibility) of interpretation derives from the fact that Postulates I to IV do not define any length scale.

More generally, a *line vortex* is defined as an asymptotic description of a vortex bundle in a limit where the bundle tends to a curve with circulation about it. Of course, such a limit excludes the simultaneous description of the velocity field in, and near, the bundle.

It is important to observe that (8.1) covers also the examples of the airfoil in two-dimensional irrotational motion mentioned in Sections 5 and 6. Indeed, if only the velocity field at distances from the airfoil that are large compared with the airfoil chord is considered, then the description may be

† Uniqueness can be proved, e.g., by complex analysis, under appropriate conditions at infinity.

based on a length scale l such that the airfoil shrinks to the origin (in the two-dimensional picture) or the x-axis (in the three-dimensional picture). For an airfoil at lift this axis has circulation, whence (8.1) follows. This example bridges the present viewpoint and that of Section 6, since the axis representing the airfoil is not in the fluid domain.

Another example of a potential vortex occurs in the most common type of hydraulic turbine or pump, where guide vanes are used to generate a motion approximated by $\phi = Ux + (2\pi)^{-1}\Gamma\theta$ in the cylindrical annulus housing the rotor blades (the x-axis is again taken as the axis of the cylinder, which is usually vertical). In this case the relevant fluid domain is $1 < R <$ const (and its axial extent is not really indefinite), so that the approximation is not an asymptotic one in the sense discussed before, but is related to the mean over θ of the velocity field.

A good deal more can be learned from the study of a pair of line vortices. It is best to start with a pair of parallel straight infinite vortices, embedded in irrotational, incompressible motion on the unbounded domain which is their complement with respect to E^3. The coordinates will be chosen so that their positions are $z = 0$, $y = \pm b$, $-\infty < x < \infty$. Since potentials are additive (Section 4), the velocity field induced by the pair is $\mathbf{v} = \text{grad} (\phi_1 + \phi_2)$ with

$$(8.3) \qquad \begin{aligned} 2\pi\phi_1 &= \Gamma_1\theta_1, & \tan \theta_1 &= z/(y - b), \\ 2\pi\phi_2 &= \Gamma_2\theta_2, & \tan \theta_2 &= z/(y + b) \end{aligned}$$

(Fig. 8.1). The motion is therefore two-dimensional, as for the potential vortex, with velocity everywhere normal to the vortices.

Consider first the case of equal and opposite circulations, i.e., $\Gamma_2 = -\Gamma_1 = -\Gamma$. The corresponding streamline pattern is shown in Fig. 8.2. Close to either one of the line vortices, the swirl induced by it dominates the motion. But at great distances from both their respective swirls largely cancel. This may be confirmed by using (8.1) to compute the velocity components referred to polar coordinates R and θ in a plane $x = $ const. They are readily found to be

$$v_R = -(2\pi)^{-1}|\Gamma|(R_1^{-1} \sin (\theta_1 - \theta) + R_2^{-1} \sin (\theta - \theta_2)),$$
$$v_\theta = (2\pi)^{-1}|\Gamma|(R_1^{-1} \cos (\theta_1 - \theta) - R_2^{-1} \cos (\theta - \theta_2))$$

in the notation of Fig. 8.1. Elementary trigonometry then shows that

$$\pi R^2 v_\theta \to |\Gamma|b \cos \theta, \qquad \pi R^2 v_R \to -|\Gamma|b \sin \theta$$

as $b/R \to 0$. This should be compared with (8.2); the vortex pair induces a velocity of magnitude

$$(8.4) \qquad |\mathbf{v}| = \pi^{-1}|\Gamma|bR^{-2}(1 + o(1)),$$

so that its field decays much more quickly with distance than that of the potential vortex; it is akin to the field of a dipole, rather than a pole.

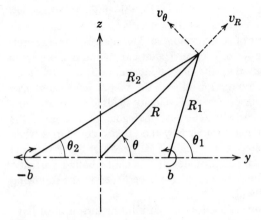

Fig. 8.1.

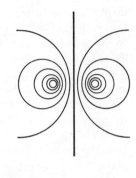

Fig. 8.2. Streamlines of the motion due to a counteroriented vortex pair.

The dominance of the swirl, close to either of the two vortices, is not quite the whole story. It is instructive to abandon the restriction to fluid domains at a fixed time, for the rest of this section, and borrow from dynamics Helmholtz' theorem (Corollary 14.3) that vortex bundles in an "ideal" fluid are convected with the fluid (i.e., the vortex bundles are fixed in Lagrangian **a**-space). Since the pair of line vortices is an asymptotic representation of a pair of vortex bundles embedded in incompressible motion, they must be similarly convected with the local fluid velocity. However, grad ϕ_1 is a pure swirl about the position of vortex 1 (at which grad ϕ_1 is undefined), and the possibility thus arises that only grad ϕ_2 might contribute to the convection of vortex 1. To prove this, consider a sequence of surfaces S_n moving with the fluid, which at $t = 0$ coincide respectively with circular cylinders $z^2 + (y - b)^2 = R_n^2$. Since grad ϕ_1 is entirely tangential to S_n at $t = 0$, it cannot contribute to the initial rate of convection of S_n for any n. Now choose the sequence so that $R_n/b \to 0$ as $n \to \infty$; then at $t = 0$, S_n tends to the same limit as the sequence of vortex tubes T_n defining the line vortex 1, and grad ϕ_2 tends to a definite vector on lim S_n. The same considerations apply to line vortex 2 (Fig. 8.1); it is convected by the velocity grad ϕ_1 induced at its position by the other vortex. The pair therefore moves in the direction of decreasing z, with speed $|\Gamma|/(4\pi b)$, at all times.

The admission is overdue, however, that not all our foregoing examples of line vortices concern asymptotic representations of the velocity field induced by vortex bundles in the fluid domain. The example of the far field of a lifting airfoil concerns a two-dimensional, irrotational motion on the domain exterior to the airfoil surface. By definition, then, this fluid domain is entirely free of vortex bundles. Nonetheless the velocity field has been seen to display a close analogy to that of a vortex bundle. This derives from the fact, shown by the Stokes' corollaries of Section 7, that circulation is more basic to the definition of vortex flux than the surface integral in Stokes' theorem. To clarify such analogies, Prandtl introduced the names *bound* and *free* vorticity. Free vorticity is the property of velocity fields defined and discussed in Section 7. Bound vorticity refers to analogous properties of the boundary $\partial\Omega_t$ of the fluid domain implied by the Stokes' corollaries when the motion has circulation. Since the boundary $\partial\Omega_t$ need not move with the fluid, Helmholtz' convection theorem is not applicable to bound vorticity. How the Stokes' corollaries may be applicable is best seen in the following example from the theory of flight.

It has already been mentioned (Section 5) that a lifting airfoil must have circulation and also that a two-dimensional flow past an airfoil of infinite span serves to approximate the flow near the center of a real wing, which must needs have finite span. It will simplify the consideration of such an aircraft wing to disregard the fuselage and engines, and to imagine the right and left wings joined at the central plane of symmetry of the aircraft (Fig. 8.3). The consideration of the wing in steady flight may then be based on the realistic premise that the intersection of the wing and the central plane of symmetry (Fig. 8.3) is a circuit $C_0 \subset \partial\Omega_t$ of circulation $\Gamma \neq 0$.

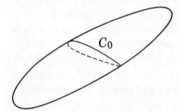

Fig. 8.3.

The reader will deduce (with surprise?) that the wing, in contrast to the airfoil, cannot be surrounded by irrotational motion. Indeed, the exterior of the wing is a simply connected domain, and so C_0 must be reducible in it.

To obtain more information, apply Stokes' theorem to the right-hand half of the wing, which is a regular surface; it shows that the component of vorticity normal to the wing surface cannot vanish everywhere on the wing. That implies both that vortex bundles must enter the fluid from this boundary and that vorticity must be defined in the wing surface. Since this surface is a part of the fluid boundary which is not convected, the vorticity in it is not free. (The full, physical meaning of this vorticity will emerge in Chapter 4.)

Moreover, the theorem applies equally to the left-hand half of the wing, so that similar vortex bundles must intersect both halves of the wing. This is not merely a consequence of symmetry. If the vortex tube of a free bundle intersects the wing in a circuit C, then C divides the wing into two parts which are surfaces to which Corollary 7.2 applies. That implies the following extension of Corollary 7.5 to the total system of free and bound vorticity: If a component W of the boundary of the fluid domain is a triangulable, closed surface, then any free vortex bundle intersecting W cannot end in W, but must re-enter the fluid domain. Observe how the conclusion that a vortex bundle cannot end is seen here to apply to the total system of free and bound vorticity together.

It will help a little to envisage the wing as that of a glider, which has a rather large span (by comparison with the wing depth) and is designed so that the circulation of the intersection of the wing surface with a plane parallel to the plane of symmetry is nearly independent of the plane chosen, except near the wing tips. Most of the free vorticity must be then "shed" near the wing tips. (When the relative humidity is high, it may cause condensation visible from the cabin as a trail of fog.) Now take the point of view of a ground-based observer. If the velocity field induced by the free vortex bundles is disregarded, they are left behind in nearly still air and, by the convection corollary, may be expected to look, from afar, rather like the trailing legs of the "horseshoe" vortex of Fig. 8.4. (On second thought, one might expect them to expand or diffuse, and it requires rather advanced dynamical considerations to explain why they are found to do so only very slowly; the contrails of high-flying aircraft, though only partly relevant, give a fairly realistic impression of the process.) It is plausible, moreover, and can be verified, that the velocity field induced by such a horseshoe vortex approaches rapidly, with increasing x/b (Fig. 8.4), the field induced by the pair of straight line vortices obtained by extending the legs of the horseshoe to $x = -\infty$ and removing its headpiece ($x = z = 0$, $|y| < b$).

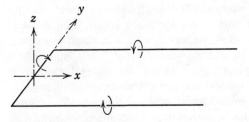

Fig. 8.4. Horseshoe vortex.

If this is accepted, it is now seen to be economically important that Corollary 7.4 applies to the bound vorticity of the wing, as well as the trailing free vorticity, and shows the vortex pair left behind to be counteroriented, like that discussed earlier (Fig. 8.1). For the far field of a co-oriented pair ($\Gamma_2 = \Gamma_1$) is readily seen from (8.3) to be the same as that of a single potential vortex of strength $\Gamma_1 + \Gamma_2$, and we could not afford to fly if that required imparting as much motion to the air as is implied by (8.2) rather than (8.4). In fact, (8.2) shows that the kinetic energy, even per unit span,

$$\frac{1}{2} \int_0^{2\pi} d\theta \int_0^\infty \rho|\mathbf{v}|^2 R \, dR,$$

of a single potential vortex of strength $\Gamma \neq 0$ is not finite.

(The definition of line vortex given above is specifically designed for the horseshoe vortex. Most line vortices must be defined by a sequence of vortex bundles B_n of strength Γ_n such that $\Gamma_n \to 0$ in a suitable manner as B_n tends to a curve. But that does not affect the preceding considerations, because the realistic applications of the potential vortex concern bounded or near-bounded domains, nor the considerations to follow, because they will be only qualitative.)

The structure of the velocity field induced by the counteroriented legs of the horseshoe explains, incidentally, why birds migrate in a V-formation. If the leading bird leaves such a pair of vortex bundles behind, it causes an upwind in the xy-plane at $|y| > b$ (Figs. 8.1 and 8.2) from which the next bird can benefit.

Corollary 7.5 shows that the free vortex bundles trailing from the wing must extend—presumably to another boundary component of the fluid domain? The horseshoe model (Fig. 8.4) shirks the issue by the device of an unbounded domain. Actually Prandtl showed the vortex system—bound and free—of a lifting wing to be closed by a "starting" vortex, which remains near the airport.

The horseshoe model of Fig. 8.4 also fails to take account of the downward convection previously deduced from Helmholtz' theorem for any pair of free, counteroriented parallel line vortices. The headpiece contributes to this downward motion of the free part of the horseshoe vortex, even if not in quite the same way as extensions of the legs to $x = -\infty$ would contribute. The proper horseshoe model for a lifting wing in steady horizontal flight is therefore such that the trailing legs asymptote, with increasing distance from the wing, to a pair of parallel, straight line vortices which are inclined to the ground and which a fixed, ground-based observer sees steadily drifting downward. This refinement of the horseshoe model accounting for inclination and drift is significant because a counteroriented pair of parallel vortices fixed in space would, on account of the symmetry of the velocity field (Fig.

8.2), induce no resultant momentum. The convection corollary is thus seen to play a role in reconciling the possibility of flight with the principle of conservation of momentum, which demands that a wing impart downward momentum to the air if it receives upward lift.

It is also instructive to apply Helmholtz' theorem to a pair of co-oriented straight parallel line vortices embedded in the irrotational motion induced by them on E^3. The potential is still given by (8.3), but with $\Gamma_2 = \Gamma_1$. The argument concerning self-induced and mutually induced velocities therefore applies, as in the case of the counteroriented pair (Fig. 8.1), but grad ϕ_2 at the position of vortex 1 and grad ϕ_1 at that of vortex 2 now have opposite directions. If the vortices are free, they therefore resolve, at all times, with angular velocity $|\Gamma|/(4\pi b^2)$ about the straight line parallel to, and midway between, them. But imagine them to be free only for $x > 0$, and bound for $x < 0$. The convection corollary then indicates that the free parts should wind and twist around each other in helical fashion like the wire strands of a spun cable!

Altogether, Helmholtz' theorem suggests that only a few, exceptional vorticity fields can have a geometrical configuration in Eulerian x-space which is invariant with respect to time, and the foregoing discussion can barely have scratched the surface of the subject of vortex motion. This is essentially because only two-dimensional motions have been considered. By (7.2), the vorticity can then have only one nonzero Cartesian component and this constraint makes such motions entirely untypical of free vortex motions.

To gain a first, qualitative impression of real vortex motions found in nature, begin with a line vortex which has, at $t = 0$, the form of a circle embedded in the irrotational incompressible motion that it induces on E^3. It is readily shown, merely by symmetry, that the induced velocity must make this vortex move, without change of shape and with constant speed, in the direction normal to its plane; and by dimensional argument, that the speed of translation is inversely proportional to the radius of the circle.

(It is also instructive to consider a pair of line vortices which are circles about the same axis in planes normal to that axis, expecially for the case in which the circulations are co-oriented. In fact, envisage a rectangular box of which the back wall is a rubber membrane and the front wall has a clean, circular hole. Fill the box with smoke and strike the back wall twice, in fairly rapid succession. Two co-oriented ring-vortices, marked by smoke, will form in succession at the rim of the hole and will travel away from the box. Again by symmetry, both will remain circular and their planes will remain parallel to the front of the box. But a moment's thought about the mutually induced velocity field will indicate that their radii may change with time, and indeed it is not obvious that the second vortex to be formed will always travel behind the first. What will happen?)

As a second example, consider a pair of line vortices, embedded in the irrotational, incompressible motion they induce on E^3, which are straight, disjoint, and perpendicular to each other, at $t = 0$ (Fig. 8.5). At that time the potential is obtained by superposing those of two potential vortices, and the mutually induced velocity (Fig. 8.5) is therefore easily computed from (8.1).

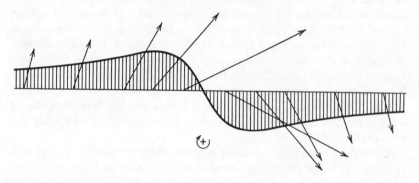

Fig. 8.5. Velocity vectors induced by a straight-line vortex at points of a straight line perpendicular to it, and rate of displacement of that line normal to itself resulting from convection with those velocities.

By the convection corollary, this field distorts the line vortices into plane curves (Fig. 8.5). But if a line vortex is not straight, then its self-induced velocity field also contributes to the convection, and for a plane line vortex this field is normal to the plane of the vortex. The convection therefore distorts these line vortices so that they do not even remain plane curves at any $t > 0$.

The reader is now invited to consider a first realistic example of natural line vortices, namely, a pair of disjoint vortices which, at $t = 0$ are circles in perpendicular planes, each passing through the center of the other. By the convection corollary and Postulates II and III, they must always remain intertwined like links of a chain. But the self-induced field propels each ring in a direction normal to its own plane, the mutually induced field (with help from the self-induced one at all $t > 0$) distorts and stretches the rings, and the smoke box example indicates that the vortices could become self-entangled in what may be hard to distinguish from knots. It becomes clear why the turbulent motion in a river full of eddies is not susceptible of detailed mathematical description.

(It will also be more understandable now why the illustrative examples in this book are drawn mostly from technology. Those from natural fluid motions tend to be too sophisticated for this Introduction. For instance, a hurricane is a striking example of a large-scale geophysical vortex. But its approximation by a potential vortex is not immediately convincing, because of the obvious importance of thermodynamic effects and vertical motion.) The attempt to follow in detail the mutually induced deformation of a pair of entwined ring vortices may make the head reel, but even a brief look at the problem will make it abundantly plausible that a salient feature of this motion—and of all natural vortex motions—is a stretching of the vortex tubes in the following sense. Since a line vortex is fixed in Lagrangian **a**-space, by Helmholtz' theorem, any segment of the vortex may be identified with a fixed interval of a Lagrangian label, and the arc length in **x**-space of the segment then becomes a well-defined function of time. Even the few qualitative facts sketched above suffice to make it clear that the mutually induced motion of a pair of entwined, initially circular line vortices is such that the arc length of many segments (even if perhaps not of all) increases with increasing time. Indeed, some initially short segments may very plausibly suffer an arbitrarily large degree of stretching in a sufficiently long time interval. (This is one of the aspects of natural fluid motion which depends decisively on the full strength of Postulate III distinguishing fluids from solids.)

Such stretching leads to an increase in the magnitude of the vorticity, as may be seen in the incompressible case from the following. Any fixed segment of a line vortex in **a**-space represents a vortex tube segment always enclosing the same mass of fluid, and hence enclosing a volume independent of time. A progressive stretching of the segment in x-space thus implies a proportional decrease of its cross-sectional area. By Kelvin's theorem (Section 14), on the other hand, a circuit fixed in Lagrangian **a**-space preserves its circulation in "ideal" fluid motion (Section 13), so that a vortex tube preserves its strength. By Stokes' theorem, therefore, an increase in arc length of a line vortex implies a directly proportional increase in the average of the normal vorticity component over any cross-section of the vortex tube represented by the line vortex. This process was first recognized by G. I. Taylor as the one primarily responsible for the production of vorticity in turbulent fluid motion.

Moreover, since some vortex tube segments must be expected to suffer an arbitrarily large degree of stretching, a local occurrence of vorticity magnitudes beyond any bound must be anticipated after a sufficient time in any natural vortex motion. At this point it is important to recall that Kelvin's theorem and its convection corollary apply only to an "ideal" fluid. For a real fluid, sufficiently large differences in vorticity will be seen in Chapters 3 and 4 to imply important viscous effects which involve a diffusion of vorticity, and thereby a bound on the vorticity differences. The study of these processes, on a statistical basis, is the object of turbulence theory [11].

9. Vortex Sheet

Consider a sequence of vortex bundles B_n progressively squashed so that $\lim B_n$ is a surface. Such a limit can furnish a useful asymptotic description of the velocity field far from a bundle by comparison with the thickness, but not with the width, of the bundle, if those dimensions are very dissimilar. But the same limit concept also arises in a quite different asymptotic sense (Section 23), which makes it a much more fundamental concept of fluid dynamics than the line vortex.

The shortest road to an adequate definition of this concept is indirect. Since it concerns a limiting motion, Postulate IV and the smoothness convention (Section 3) need not apply. Just as the velocity is not well defined at the position of a line vortex, so it must be anticipated that the velocity, or some of its derivatives, may tend to limits, if any, which depend on the direction of approach to the surface which is $\lim B_n$. It then becomes necessary to distinguish between the two sides of such a surface, in a manner similar to the natural distinction between the two sides of a thin plate or wing immersed in fluid. However, as an asymptotic representation of a *free* vortex bundle, the surface which is $\lim B_n$ should be a subset of the fluid domain Ω_t. That suggests defining a *discontinuity surface* as an orientable surface $S \subset \Omega_t$ at which the smoothness convention is not satisfied, but to either side of which the velocity, density (and stress), and their space derivatives have well-defined extensions by continuity.

Definition. *Free vortex sheet* means a regular, discontinuity surface in Ω_t across which the tangential velocity, but not the normal velocity component, is discontinuous.

The connection with vorticity is clarified by the following. Let the two sides of a vortex sheet S be distinguished by subscripts $+$ and $-$, let $\mathbf{n}$ denote the unit normal on S toward the $+$-side, and define

$$(9.1) \qquad \boldsymbol{\Omega} = \mathbf{n} \wedge (\mathbf{v}_+ - \mathbf{v}_-),$$

where $\wedge$ denotes the vector product. This is a tangential vector field on S, and since $\mathbf{v}_+ - \mathbf{v}_-$ is also tangential to S, the rule on the triple product gives

$$(9.2) \qquad \mathbf{v}_+ - \mathbf{v}_- = \boldsymbol{\Omega} \wedge \mathbf{n}.$$

For any path $P \subset S$, let s and $\mathbf{s}$ denote respectively the arc length and unit tangent vector in the sense of the orientation of P. Then with $\boldsymbol{\nu} = \mathbf{n} \wedge \mathbf{s}$,

$$(9.3) \qquad \int_P (\mathbf{v}_+ - \mathbf{v}_-) \cdot d\mathbf{x} = \int_P \boldsymbol{\Omega} \cdot \boldsymbol{\nu} \, ds,$$

by (9.2), and this is a limiting form of Stokes' theorem (Section 7) in which the left-hand side represents the circulation of the circuit $P_+ - P_-$ (Fig. 9.1), and the right-hand side, the flux of $\boldsymbol{\Omega}$ through the limit of a surface which spans the circuit with surface normal $\boldsymbol{\nu}$ related by the right-hand screw rule to the orientation of the circuit.

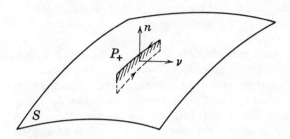

Fig. 9.1.

The vector field $\boldsymbol{\Omega}$ on S is accordingly called the vortex strength (per unit area), and its trajectories are called the vortex lines of the sheet. Moreover, any open set $\delta \subset S$ may be used to define a vortex strip of the sheet as the union of the points of all the vortex lines intersecting it. Since no vortex lines can cross the boundary of such a strip, it is seen to be a limit of a sequence of vortex bundles, and Corollaries 7.4 and 7.5 apply to it in the sense of (9.3).

The simplest example is a vortex sheet S occupying the strip $|x| < \infty$, $|y| < b, z = 0$ (Fig. 9.2), with strength $\boldsymbol{\Omega} = \Omega(y)\mathbf{i}$ ($\mathbf{i}, \mathbf{j}$, and $\mathbf{k}$ will denote

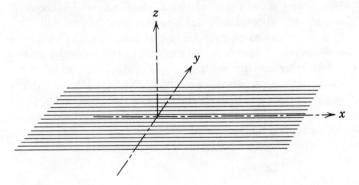

Fig. 9.2.

respectively the unit vectors in the directions of increasing x, y, and z), embedded in incompressible, irrotational motion on an unbounded domain. The corresponding velocity field is clearly two-dimensional, and is plausibly obtainable from (8.1) and (8.2) by superposition. At a point $(x, y, 0)$ on the sheet itself this gives readily

$$\mathbf{v} = -(2\pi)^{-1}\mathbf{k} \int_{-b}^{b} (y - y')^{-1}\Omega(y') \, dy',$$

but the integral is singular, and in fact the result is false because inadequate account has been taken of the partial breakdown of Postulate IV at the vortex sheet.

It is necessary to recall the strict definition, at the beginning of this section, of the two sides of the vortex sheet as distinct subsets of $\partial\Omega_t$ and to calculate the respective velocities on them as distinct continuous extensions from Ω_t. Some consideration of the velocity field on Ω_t, rather than just on the vortex sheet, is therefore needed.[†] That field is two-dimensional, with components 0, v, and w, and by (6.3) it must have the property that $v - iw$ is an analytic function of $\zeta = y + iz$ on Ω_t. The function

$$(9.4) \qquad v - iw = \frac{1}{2\pi i} \int_{-b}^{b} (\zeta - y')^{-1}\Omega(y') \, dy'$$

suggested by (8.2) has that property, if $\Omega(y)$ is Holder-continuous on $(-b, b)$ (i.e., there exist $M, \alpha > 0$, such that $|y| < b$ and $|y'| < b$ imply $|\Omega(y) - \Omega(y')| < M|y - y'|^{\alpha}$). On $\partial\Omega_t$, that is, for $\zeta = y \pm i0$, with $|y| < b$, the Plemelj Formulae [12] give the boundary values of (9.4) as

$$(9.5) \qquad v - iw = \frac{1}{2\pi i} CP \int_{-b}^{b} (y - y')^{-1}\Omega(y') \, dy' \mp \tfrac{1}{2}\Omega(y),$$

where CP denotes the Cauchy Principal Value. The function (9.4) therefore possesses precisely the discontinuity defining the vortex sheet, and since $|v - iw| \to 0$ as $|\zeta| \to \infty$ and also for every ζ as $\text{lub}_{|y| \le b} |\Omega(y)| \to 0$, (9.4) is called the velocity field due to the vortex sheet. The Plemelj Formula (9.5) conveniently represents the velocity at the vortex sheet itself as the sum of a discontinuous, tangential velocity $\mp\tfrac{1}{2}\Omega\mathbf{j} = \pm\tfrac{1}{2}\mathbf{\Omega} \wedge \mathbf{k}$ and a continuous, normal velocity at $(x, y, 0)$ due to $\Omega(y')$ with $y' \ne y$. The latter is the velocity with which the vortex sheet is convected in ideal fluid motion, at the time under consideration, by Helmholtz' theorem.

An interesting particular case is that for which

$$(9.6) \qquad \Omega(y) = \gamma y b^{-1}(b^2 - y^2)^{-\frac{1}{2}}$$

with constant γ (the streamline pattern in the yz-plane of the corresponding velocity field (9.4) is shown in Fig. 9.3, which should be compared with Fig. 8.2). The principal value of the integral in (9.5) can then be shown to be $\pi\gamma/b$, independent even of y, and this free vortex sheet is therefore convected

---

† A reader unfamiliar with the analytical technique appropriate here will find an incomplete comprehension of the next few lines no handicap in understanding what follows thereafter.

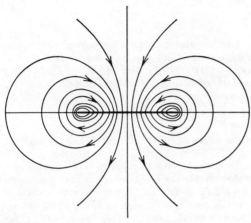

Fig. 9.3.

in the z-direction without change of shape, like a plank—a property which Prandtl and Munk used to show that this particular vortex sheet is characteristic of the shed vorticity of a wing of minimum "induced drag" (meaning the part of the drag attributable to the kinetic energy of the air motion left behind by the wing in ideal fluid motion). But the same property also makes it an exceptional free vortex sheet. For suppose that only the part $y > 0$ of it were present; then Ω would have the same direction and sense at every point of the sheet, and the discussion of the co-oriented vortex pair (Section 8) indicates that such a sheet must roll up at $t > 0$ if it was flat at $t = 0$. Indeed, the Prandtl-Munk sheet is known to be unstable as a free vortex sheet in the sense that an initially flat straight sheet of strength distribution differing arbitrarily little from (9.6) will roll up into a shape with cross-section forming a pair, or pairs, of spirals.†

It is also instructive to note the mathematical awkwardness occurring at the edge of a vortex sheet, unless $|\Omega| \to 0$ as the edge is approached; $\mathbf{v}$ then fails to possess a continuous extension to all of the sheet. For instance, even the Cauchy Principal Value of $\int_{-b}^{b} (b - y')^{-1}\Omega(y')\, dy'$ fails to exist unless $\Omega(b) = 0$, and in the Prandtl-Munk example also $|\mathbf{v}_+ - \mathbf{v}_-| \to \infty$ as $|y| \uparrow b$. The vortex sheet is therefore only partially successful as an asymptotic representation of a wide, thin vortex bundle for the purpose of describing the induced velocity field at distances from the bundle that are large compared with its thickness, but not compared with its width. It is not generally an asymptotic approximation of *uniform* validity in such a sense.

† This, rather than the shedding of more vorticity from the wing tips than the wing center, is the decisive reason why the horseshoe vortex is a good asymptotic approximation for the vortex system of a wing (Section 8).

Appendix 9

The translation rules of Section 3 are not always adequate for limiting motions with discontinuity surfaces. To extend them, consider any open bounded set $D(t) \subset E^3$ of points $x(t)$ such that $dx/dt = V(x, t) \in C^1(D) \cap C(\bar{D})$. Then if $f(x, t) \in C^1(\bar{D})$, the f content of D is $\int_{D(t)} f \, dV = F(t)$, and its rate of change is obtained from the Convection Theorem and Corollary 4.1, with V in the place of v, as

$$(9.7) \qquad \frac{dF}{dt} = \int_{D(t)} \frac{\partial f}{\partial t} \, dV + \int_{\partial D} f V \cdot n \, dS.$$

Thus dF/dt at time t depends on the values of $x(t)$ at other times only through the normal velocity $V \cdot n$ of ∂D at time t.

Next consider a fluid domain Ω_t (with regular boundary surface $\partial \Omega_t$) separated by a regular surface $S \subset \Omega_t$ into two subdomains Ω_+ and Ω_-. Let n denote the outward unit normal on $\partial \Omega_t$ and that directed toward the $+$ side, on S; let $V \cdot n$ denote the normal velocity of the moving surface S itself (Section 3) in the sense of n; and let v_+ and v_- denote the respective unit outward normals on $\partial \Omega_+$ and $\partial \Omega_-$. Then if $f(x, t) \in C^1(\bar{\Omega}_+) \cap C^1(\bar{\Omega}_-)$, (9.7) gives

$$\frac{d}{dt} \int_{\Omega_*} f \, dV = \int_{\Omega_*} \frac{\partial t}{\partial t} \, dV + \int_{\partial \Omega_*} f U \cdot v_* \, dS, \qquad * = + \quad \text{or} \quad -,$$

where $U \cdot v_+ = -V \cdot n$ on $S \cap \partial \Omega_+$, since $v_+ = -n$ there; $U \cdot v_- = V \cdot n$ on $S \cap \partial \Omega_-$; and $U \cdot v_* = v \cdot n$ on $\partial \Omega_* \cap \partial \Omega_t$, since $V = v$ on $\partial \Omega_t$. Adding yields, since Ω_t is the domain of an individual body of fluid,

$$(9.8) \quad \frac{d}{dt} \int_{\Omega_t} f \, dV = \frac{D}{Dt} \int_{\Omega_t} f \, dV = \int_{\Omega_t} \frac{\partial f}{\partial t} \, dV + \int_{\partial \Omega_t} f v \cdot n \, dS - \int_S [f] V \cdot n \, dS,$$

where [ ] denotes the saltus across S, from the $-$ to the $+$ side, of the quantity in brackets. It should be stressed that this extension of (3.6) holds for surfaces in Ω_t; it does not necessarily hold for immersed boundary surfaces which are an asymptotic representation of physical boundaries and may therefore be a source of the quantity f.

Problem 9.1. Show that $\rho(v - V) \cdot n$ is continuous across a regular discontinuity surface S if its normal velocity $V \cdot n$ is a continuous function on S.

CHAPTER 2

Momentum Principle and Ideal Fluid

10. Conservation of Linear Momentum

Postulate VI (Momentum Principle). A vector field **f** and a tensor field with components p_{ij} are defined on the closure of any fluid domain Ω_t with regular boundary surface $\partial\Omega_t$ so that

$$(10.1) \qquad \frac{D}{Dt} \int_{\Omega_t} \rho v_i \, dV = \int_{\Omega_t} \rho f_i \, dV + \int_{\partial\Omega_t} p_{ij} n_j \, dS, \qquad i = 1, 2, 3,$$

where **n** is the unit outward normal on $\partial\Omega_t$.

The left-hand side of (10.1) represents the rate of change of the linear momentum of the body of fluid which occupies the domain Ω_t, and the right-hand side therefore represents the resultant external force exerted on this body of fluid. The postulate states that external force can be exerted on the fluid in two ways: by a *stress* acting on the surface of the fluid and by a *body force* $\rho\mathbf{f}$ acting directly on each volume element. The most common body forces are gravity, centrifugal and Coriolis forces (to be considered only in Chapter 5), and electromagnetic forces (not considered in this book). Since the postulate applies also to every subdomain of a fluid domain, $p_{ij}n_j$ represents the *i*-component of force per unit area exerted across an area element with "outward" unit normal **n** by the fluid on the "outer" side of the area element upon the fluid on its "inner" side.

It should be emphasized that this postulate is more fundamental than those of Sections 1 and 2, in the sense that Postulates V, VI, and VIII (in contrast to Postulates I to IV) hold for all of continuum mechanics.

It may also be noted that the momentum principle does not require the pointwise definition of f_i and p_{ij}. Indeed, the values of these quantities (and of the density) at a point are not directly observable, and only the integrals occurring in Postulates V and VI have strict physical meaning. As noted in Section 3, however, ρ and p_{ij} are interpretable as statistical averages over mechanical properties of molecules, and the smoothness convention may therefore be taken to apply to them in the sense discussed in Section 3. Moreover, there does not appear to be any real loss of physical generality in taking that convention to apply also to the body force field.

Corollary 4.1 may be used to translate the momentum principle into the Eulerian form

$$(10.2) \qquad \int_{\Omega_t} \rho \frac{Dv_i}{Dt} \, dV = \int_{\Omega_t} \rho f_i \, dV + \int_{\partial \Omega_t} p_{ij} n_j \, dS.$$

Alternatively, Corollary 3.1 permits us to write

$$(10.3) \qquad \frac{D}{Dt} \int_{\Omega_t} \rho v_i \, dV = \frac{\partial}{\partial t} \int_{\Omega_1} \rho v_i \, dV + \int_{\partial \Omega_1} \rho v_i \mathbf{v} \cdot \mathbf{n} \, dS,$$

where Ω_1 denotes the fixed set in E^3 coinciding with Ω_t at the time t. The last integral represents the net rate of outflow of linear momentum through the boundary of Ω_1. This quantity therefore represents the difference between the rate of change of the linear momentum of a body of fluid and the rate of change of linear fluid momentum in the fixed domain Ω_1 momentarily occupied by that body of fluid. Another, purely Eulerian, form of the momentum principle is therefore

$$(10.4) \qquad \frac{\partial}{\partial t} \int_{\Omega} \rho v_i \, dV + \int_{\partial \Omega} \rho v_i \mathbf{v} \cdot \mathbf{n} \, dS = \int_{\Omega} \rho f_i \, dV + \int_{\partial \Omega} p_{ij} n_j \, dS,$$

where Ω is any domain in E^3 occupied by fluid whose boundary $\partial \Omega$ is a regular surface with unit outward normal $\mathbf{n}$, and this is the form of the Principle most frequently used in fluid dynamics.

The mathematical comment may be added that the basic, Lagrangian form (10.1) of the momentum principle is linear in the velocity $\mathbf{v}$, if (as will emerge to be the usual case) the body force and stress tensor do not possess a direct, nonlinear dependence on the velocity. The nonlinearity of the form (10.4) is shown by (10.3) to be due to the translation from the Lagrangian to the Eulerian representation. The nonlinear "convective" (because it arises from the Eulerian translation of the convective derivative D/Dt) or "inertia" term $\int \rho v_i v_j n_j \, dS$ in (10.4) is seen to have a mathematical form similar to that of the resultant force on the boundary, and $\rho v_i v_j$ is sometimes called a stress. Indeed, it represents a momentum transfer across fixed surfaces by the fluid motion analogous to the molecular momentum transfer (Section 18) represented by the stress tensor.

The comment just made is not meant to imply that the nonlinearity of (10.4) is artificial. In the Lagrangian representation a corresponding nonlinearity arises from the implicit involvement of the Jacobian $\det (\partial x_i / \partial a_k)$ in the statement of mass conservation. (In the incompressible case (4.5) shows that nonlinearity to be removed by translation to the Eulerian representation.)

The presence, in one form or another, of the nonlinearity is deeply rooted in our basic notions on fluid motion, and the nonlinearity is largely responsible for the fact that in 200 years of active research few detailed results have been obtained that are not of an asymptotic nature in one sense or another. The nonlinearity also indicates that the smoothness convention (or any similar regularity postulate permitting the conversion of the conservation principles into differential equations) may be inconsistent with Postulates V and VI, unless Postulate I is interpreted strictly.

The use of the momentum principle for any detailed predictions of fluid motions depends on specification of the body force and stress tensor. A property of the stress generally accepted to be common to all fluids is that in a fluid at rest $p_{ij} = -p(\mathbf{x})\delta_{ij}$, where δ_{ij} is Kronecker's symbol signifying unity if $i = j$, and zero otherwise. Now, it is a physical axiom that fluid dynamics, like all of "classical" physics, is invariant under Galilean transformations, that is, the physical laws have the same form in any two reference frames moving relative to each other with a fixed velocity. The Postulates I to VI are so invariant (and those to follow later will also be seen to be), if differences in p_{ij} are so invariant, and this is therefore a basic condition to be imposed on any specification of the stress. It follows that the fluid property under discussion must extend to *fluid in uniform motion*, meaning any fluid domain Ω_t on which $\mathbf{v}(\mathbf{x}, t) \equiv$ const over an open time interval. It may thus be formulated as

Postulate VII. In a fluid in uniform motion $p_{ij} = -p(\mathbf{x}, t)\delta_{ij}$.

More precisely, this means that there is a scalar field $p(\mathbf{x}, t)$, the *pressure*, such that $|p_{ij} + p\delta_{ij}|$ is arbitrarily small for every i and j at an interior point $\mathbf{x}_0$ of a fluid domain at time t_0, if there is a neighborhood N of $\mathbf{x}_0$ and an open time interval I about t_0 such that $\text{lub}_{\mathbf{x}\in N, t\in I} |\mathbf{v}(\mathbf{x}, t) - \mathbf{v}(\mathbf{x}_0, t_0)|$ is sufficiently small.

This suggests that we write $p_{ij} = -p\delta_{ij} + \tau_{ij}$ for the general case of nonuniform fluid motion, and if p is identified here with the thermodynamic pressure (Sections 33 and 38), then τ_{ij} is called the *viscous* stress tensor. Now, it will be helpful to divide the difficulties by postponing any thermodynamical considerations to the last chapter. For most fluids such considerations may be avoided by restricting attention to incompressible motion (Section 4), and this will be done in Sections 13, 14, and 19 to 32. For the Newtonian fluid, with which this book is primarily concerned, the trace τ_{kk} of the viscous stress tensor will be seen in Section 17 to be proportional to div $\mathbf{v}$, which vanishes for incompressible motion, by (4.5), and we may then use

(10.5) $$p_{ij} = -p\delta_{ij} + \tau_{ij}, \qquad \tau_{11} + \tau_{22} + \tau_{33} = 0$$

as a definition of the pressure in nonuniform motion. In any case τ_{ij} contains all of the off-diagonal components of p_{ij}, which represent the shear stresses, and for Chapters 2 to 4 it will help the reader to think of the viscous stress τ_{ij} as essentially representing the shear stress.

The variation of the pressure $p(\mathbf{x}, t)$ in a fluid in uniform motion is determined by the momentum principle. By Postulates VI and VII and Corollary 4.1,

$$\int_{\Omega_t} \rho \mathbf{f} \, dV = \int_{\partial\Omega_t} p\mathbf{n} \, dS$$

for every subdomain Ω_t with regular boundary surface, and since it has already been seen that $\rho\mathbf{f}$ and grad p can be regarded as continuous in such a motion, the Divergence theorem leads to the hydrostatic law that

(10.6) $$\text{grad } p = \rho\mathbf{f}$$

in a fluid in uniform motion.

There is a class of body force fields, including that of gravity, such that $\mathbf{f}$ is the gradient of a (single-valued) potential $\Phi(\mathbf{x})$. If, in addition, the motion is incompressible (Section 4) or ρ is a function only of Φ (as in an atmosphere in equilibrium), the body force may conveniently be eliminated from explicit consideration by introducing the *hydrodynamic pressure*

(10.7) $$p' = p - \int^{\mathbf{x}} \rho(\text{grad } \Phi) \cdot d\mathbf{x}.$$

Substitution in (10.1) shows the body force integral to be absorbed in the stress integral if p_{ij} is interpreted as $-p'\delta_{ij} + \tau_{ij}$. Since the remainder of this chapter, and Chapters 3 and 4, will be concerned primarily with motions for which these conditions are satisfied, the convention (common in the literature) will be adopted that pressure is always understood to mean hydrodynamic pressure

$$p = -\tfrac{1}{3}(p_{11} + p_{22} + p_{33}) - \int^{\mathbf{x}} \rho(\text{grad } \Phi) \cdot d\mathbf{x}.$$

(The slightly indeterminate notation reflects the invariance of (10.1) under addition of a constant multiple of δ_{ij} to p_{ij}.) Accordingly the body force integral will be omitted in the momentum principle, and the hydrostatic law (10.6) will be applied in the form of

Lemma 10.1. $p(\mathbf{x}, t) = $ const in a fluid in uniform motion.

Problem 10.1. A bounded solid body fixed in space is immersed in a fluid in steady motion without body force field. To compute the force $\mathbf{F}$ exerted by the fluid on the body, the most obvious approach is to express $\mathbf{F}$ as an integral of the stress over the body surface, but more often than not this is inconvenient. (In the case of a bird, for instance, precise information on the stress on the body surface is virtually unobtainable, and even for technological bodies difficulties arise when the surface is not smooth. It is not uncommon, therefore, that the quality of the readily available information on the fluid motion increases with distance from the body.) Let S denote any fixed, closed, regular surface in the fluid such that the body lies in the subset of E^3 interior to S. Express $\mathbf{F}$ as an integral over S of quantities defined on S by the fluid motion.

Appendix 10

The Eulerian forms (10.2) to (10.4) of the momentum principle, being based on the translation rules of section 3, are not necessarily valid for limiting fluid motions. If Ω_t is a fluid domain separated by a regular, discontinuity surface S into subdomains on which the smoothness convention is valid, (10.1) and (9.7) give

$$\int_{\Omega_t} \frac{\partial}{\partial t} (\rho v_i) \, dV + \int_{\partial \Omega_t} \rho v_i \mathbf{v} \cdot \mathbf{n} \, dS - \int_S [\rho v_i] \mathbf{V} \cdot n \, dS$$

$$= \int_{\Omega_t} \rho f_i \, dV + \int_{\partial \Omega_t} p_{ij} n_j \, dS, \qquad i = 1, 2, 3,$$

where $\mathbf{n}$ denotes the unit outward normal on $\partial \Omega_t$ (assumed to be a regular surface) and that directed toward the $+$ side, on S, [ ] denotes the saltus across S, and $\mathbf{V} \cdot \mathbf{n}$ represents the normal velocity of S. This relation may be applied to a nesting sequence of subdomains such that the subdomain boundaries tend to a surface coinciding with the two sides of S. Since the subdomain volumes then tend to zero, passage to the limit gives

$$\int_S [\rho v_i \mathbf{v}] \cdot \mathbf{n} \, dS - \int_S [\rho v_i] \mathbf{V} \cdot \mathbf{n} \, dS = \int_S [p_{ij}] n_j \, dS.$$

This holds equally for every open subset of S and therefore implies

Lemma 10.2. $p_{ij} n_j - \rho v_i (\mathbf{v} - \mathbf{V}) \cdot \mathbf{n}$ is continuous across a regular, discontinuity surface S with unit normal $\mathbf{n}$ if its normal velocity $\mathbf{V} \cdot \mathbf{n}$ is continuous on S.

A free vortex sheet in ideal fluid will be seen from Corollary 14.3 to be a surface moving with the fluid, so that (3.4) implies

Corollary 10.2. The stress $p_{ij} n_j$ ($i = 1, 2,$ or 3) is continuous across a free vortex sheet with unit normal $\mathbf{n}$.

By contrast, the lemma shows that a bound vortex sheet must generally be supported in position by an externally applied stress.

11. Mixing and Lift

The nature of the conservation principles for mass and momentum is best appreciated by noting their similarity to the basic laws of thermodynamics. Like those laws, the conservation principles are nonspecific, offering a framework for the description of physical processes rather than quantitative predictions. But if a small amount of observational information, or a modest set of physical assumptions, be added, the conservation principles reveal themselves as remarkable tools for deducing a wealth of precise information. In particular, the most important information may turn out to depend to a surprisingly small degree on the details of the process and may then be uncovered by the conservation principles without any analysis of those details.

This may be illustrated by two examples of Prandtl [13]. The first concerns two straight pipes, with common axis, filled with fluid, which issues steadily from the narrower into the wider pipe (Fig. 11.1). The pipes are joined by a solid annular disc, and the mouth of the narrower pipe has a sharp edge. If the fluid issues from it with any appreciable velocity, it is a matter of everyday experience that it issues as a jet, whose boundary is a surface across which the tangential velocity changes rather sharply—and hence, a surface resembling a vortex sheet. The physical process is one of mixing between the fluid within and without the jet, in which the jet diffuses and gradually widens to fill the larger pipe (Fig. 11.1). It is a complicated physical process, and no attempt will be made here to discuss it in any detail. Instead the analysis of the motion will be based on the assumptions (i) that the part of the jet boundary close to the mouth of the narrower pipe may be approximated by a vortex sheet, and (ii) that the flow may be approximated as a uniform one in the narrower pipe and again in the larger pipe at a sufficient distance from the mouth of the narrower one—as is not implausible and is experimentally confirmed for fluids of small viscosity.

Let the x_1-axis be taken along the common axis of the pipes, with x_1 measured from the mouth of the narrower pipe in the direction of the fluid stream (Fig. 11.1). Let Ω denote the fluid domain bounded by the plane

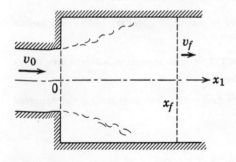

Fig. 11.1.

$x_1 = 0$, the wall of the larger pipe, and a plane $x_1 = \text{const} = x_f > 0$ at a sufficient distance from $x_1 = 0$. Let A_0 denote the area of cross-section of the narrower pipe, and A_f, that of the larger one. As in Section 4, conservation of mass implies

$$(11.1) \qquad \rho_0 v_0 A_0 = \rho_f v_f A_f,$$

where $\rho_0 v_0$ stands for the average of ρv_1 over the cross-section of the narrower pipe at $x = 0$, and $\rho_f v_f$, for the average of the same quantity over the cross-section of the larger pipe at $x = x_f$. Since the pipe wall is an impermeable boundary, the net rate of outflow of x-component of linear momentum through the boundary of Ω is

$$(11.2) \qquad \int_{\partial \Omega} \rho v_1 \mathbf{v} \cdot \mathbf{n} \, dS = \rho_f v_f{}^2 A_f - \rho_0 v_0{}^2 A_0,$$

where $\rho_f v_f{}^2$ and $\rho_0 v_0{}^2$ now stand for the respective averages of $\rho v_1{}^2$. Assumption (ii) permits us to ignore the distinction between the interpretations of the symbols in (11.1) and (11.2), so that

$$(11.3) \qquad \int_{\partial \Omega} \rho v_1 \mathbf{v} \cdot \mathbf{n} \, dS = \rho_f v_f A_f (v_f - v_0).$$

Assumption (ii) also implies, by Postulate VII and Lemma 10.1, that $-p_{1j} n_j$ is constant over the part of $\partial \Omega$ in the plane $x = x_f$ and equal to the (hydrodynamic) pressure p_f there. Similarly, $p_{1j} n_j = \text{const} = p_0$ over the part of $\partial \Omega$ which forms the mouth of the narrower pipe.

Outside that pipe, the part of the fluid domain bounded by the solid walls and the jet boundary is a region in which at most a circulating motion of small velocity can plausibly be expected. It follows from Postulate VII and the hydrostatic law that the stress tensor may be approximated there by $-p \delta_{ij}$, and the pressure p, by a constant. But this constant must also be p_0, by assumption (i) and Corollary 10.2.

The momentum principle therefore implies, by (10.4) and (11.3),

$$(11.4) \qquad \rho_f v_f A_f (v_f - v_0) = (p_0 - p_f) A_f - T,$$

if T denotes the resultant shear force exerted on the fluid in Ω by the wall of the wider pipe. If x_f is not chosen too large, $|T|/(A_f |p_0 - p_f|)$ is found experimentally to be a small quantity for fluids of small viscosity, and will be neglected for consistency with the earlier assumptions.

For incompressible motion, where $\rho_f = \rho_0$, (11.1) and (11.4) become, respectively,

$$(11.5) \qquad v_0/v_f = A_f/A_0,$$

$$(11.6) \qquad p_f - p_0 = \rho v_f{}^2 [(A_f/A_0) - 1].$$

The foregoing discussion may not resemble conventional mathematics, since it does not proceed from a set of differential equations with boundary conditions. But in fact it is an analysis deducing a considerable amount of information rigorously from much weaker assumptions than are required for formulating differential equations approximating the physical process. To appreciate the character of the information gained, observe that (11.5) and (11.6) permit us to obtain both the velocities v_0 and v_f from a measurement of only one pressure difference. Alternatively, if conditions are known in the narrower pipe, approximate predictions of the average pressure and velocity at $x = x_f$ have been obtained for any fluid of small viscosity in the absence of almost any understanding of the fluid motion, and experiment shows these approximations to be rather close. (The actual flow encountered in an experiment is usually steady only in an average sense, and the quantities considered must then be interpreted as averages also over time; the distinction between the time average of $\rho v_1{}^2$ and the product of those of v_1 and ρv_1 usually turns out also to have a small effect on the result.) It is instructive to notice that a rather different result is obtained from a similar application of the momentum principle to the case in which the transition from the narrower to the wider pipe is smooth (Fig. 4.2) rather than abrupt, and the stream forms no jet.

An even more remarkable example, on account of the great economy of assumptions, is Prandtl's derivation of Joukowski's theorem on the lift of an airfoil in steady flow. One of the main difficulties here lies in the convincing formulation of the conditions "at infinity," and Prandtl avoided this by considering a cascade, that is, a countably infinite set of airfoils of identical shape placed in a regular row (Fig. 11.2). The airfoils are solid cylinders with impermeable surface and generators parallel to the z-axis; the y-axis is taken parallel to the row (Fig. 11.2).

Contemplation of an experimental realization of the fluid motion indicates that the conditions far upstream are at the disposal of the experimenter, and it will be specified that the motion tends to a uniform one, as $x \to -\infty$ for all y, of density ρ_1 and velocity $\mathbf{v}_1$ parallel to the plane of Fig. 11.2. It is then highly plausible that the whole flow is two-dimensional, with velocity parallel to the xy-plane, and periodic in y, with period equal to the spacing d (Fig. 11.2), and this will be assumed. Since the flow is steady, mass cannot accumulate in any fixed region, which suggests that a streamline S can be traced across all planes $x = $ const, and this will be proved below in the more general context of Theorem 13.1. While it is not necessary, it will simplify the ideas to take S to be disjoint from the airfoil surfaces (Fig. 11.2). Finally, it will be assumed—as confirmed by experiment—that the flow tends to a uniform one also as $x \to +\infty$, with density ρ_2 and velocity $\mathbf{v}_2$, say.

The momentum principle will now be applied to a fluid domain Ω bounded by one of the airfoils, two planes normal to the z-axis at unit distance apart, two planes $x = $ const, and two cylindrical surfaces, with generators in the z-direction, formed by two replicas of S at a distance d (measured in the

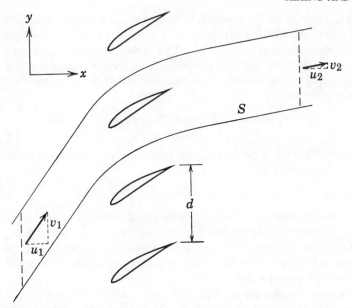

Fig. 11.2.

y-direction) apart (Fig. 11.2). The planes $x = $ const are taken at $x = x_1 < 0$ and $x = x_2 > 0$, with $|x_1|$ and x_2 sufficiently large; in fact, we consider a sequence of such domains differing only in x_1 and x_2 and such that $|x_1|$ and x_2 tend to infinity. Let the x- and y-components of $\mathbf{v}$ be respectively u and v. Since $\mathbf{v} \cdot \mathbf{n} = 0$ at any point of $\partial\Omega$ at which neither $x = x_1$ nor $x = x_2$, the mass conservation statement (4.2) implies in the limit, on account of the steadiness and periodicity of the motion,

$$\rho_2 u_2 = \rho_1 u_1.$$

Similarly, the net rates of outflow across $\partial\Omega$ of x- and y-momentum tend respectively to

$$\rho_2 u_2 d \cdot u_2 - \rho_1 u_1{}^2 d = \rho_1 u_1 d(u_2 - u_1)$$

and $$\rho u_2 d \cdot v_2 - \rho_1 u_1 v_1 d = \rho_1 u_1 d(v_2 - v_1),$$

and $(v_2 - v_1)d \equiv \Gamma$ is, on account of the periodicity, the circulation (taken anticlockwise) of the circuit which is the intersection of $\partial\Omega$ with a plane $z = $ const.

The periodicity also implies that the stresses acting on the fluid in Ω across the curved parts of $\partial\Omega$ have no resultant force; nor do those on the faces of $\partial\Omega$ parallel to the xy-plane. Those acting across the faces $x = x_1$ and $x = x_2$ result, in the limit, in a force with components $(p_1 d - p_2 d, 0)$, by Postulate

VII, where p_i, $i = 1, 2$, are the (hydrodynamic) pressures at $x = x_i$, which are constants, by Lemma 10.1. Finally, the stresses across the airfoil surface have a resultant force with components $(-X, -Y)$, if X and Y are the components of the *aerodynamic force*, per unit span in the z-direction, exerted by the fluid on an airfoil of the cascade. The statement (10.4) of the momentum principle therefore implies.

$$(p_1 - p_2)d - X = \rho_1 u_1 d(u_2 - u_1),$$
$$- Y = \rho_1 u_1 \Gamma.$$

For incompressible flow in particular these statements of mass and momentum conservation reduce to

$$u_1 = u_2 = u, \qquad X = (p_1 - p_2)d, \qquad Y = -\rho u \Gamma.$$

These relations represent a remarkable amount of information, obtained without any consideration of the flow pattern near the airfoils. For instance, with the conditions far upstream considered known, the aerodynamic force has been shown to be deducible from a measurement only of the pressure and velocity at a single point far downstream.

It is more usual to calculate the drag D and lift L defined as the respective components of the aerodynamic force in the direction of, and normal to, the mean velocity $\mathbf{V}$ with the components u and $\bar{v} = (v_1 + v_2)/2$ (Fig. 11.3).

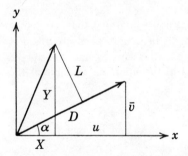

Fig. 11.3.

This gives $D = X \cos \alpha + Y \sin \alpha$, $L = Y \cos \alpha - X \sin \alpha$, with $\cos \alpha = u/V$, $\sin \alpha = \bar{v}/V$, where $V = |\mathbf{V}|$ (Fig. 11.3). Substitution of the values of X and Y previously found for the incompressible case gives

(11.7)
$$VD = ud[p_1 - p_2 + \tfrac{1}{2}\rho(|\mathbf{v}_1|^2 - |\mathbf{v}_2|^2)],$$
$$VL = -\rho u^2 \Gamma - \bar{v}(p_1 - p_2)d,$$

whence

(11.8)
$$L = -\rho V\Gamma - (\bar{v}/u)D.$$

This last relation does not depend explicitly on the spacing d (Fig. 11.2), and remains valid as $d \to \infty$ with Γ and D fixed. But in that limit, $v_2 - v_1 \to 0$, by the definition of Γ, and so (11.8) applies to the isolated airfoil in an unbounded stream of velocity $\mathbf{V}$ far upstream and far downstream; and after a change of Galilean observer, also to the isolated airfoil moving at constant flight velocity $-\mathbf{V}$ into unbounded fluid at rest far ahead and far astern. Joukowski's theorem is the special case of (11.8) resulting from an additional assumption (to be introduced in Section 13) implying that $p + \frac{1}{2}\rho|\mathbf{v}|^2$ is constant along S and hence $D = 0$.

Problem 11.1. A vessel is formed by the solid cylinder $x^2 + y^2 = R^2$, $|z| < \infty$, into which a slit has been cut from $x = |(R^2 - b^2)^{\frac{1}{2}}|$, $y = b$, to $x = |(R^2 - b^2)^{\frac{1}{2}}|$, $y = -b$ (Fig. 11.4); two plane, solid, impermeable sheets $y = \pm b$, $0 < x < |(R^2 - b^2)^{\frac{1}{2}}|$, $|z| < \infty$, are welded to the edges of the slit. The vessel is filled with inviscid, incompressible liquid and the slit leads to an unbounded domain filled with air. Liquid is now allowed to seep into the vessel through the cylinder at a rate independent of position on the cylinder, and a steady, two-dimensional, irrotational flow is assumed to develop. Assume also that the liquid separates from the vessel walls at $x = 0$, $y = \pm b$ (Fig. 11.4), to form a jet issuing between the plane sheets, the jet boundaries being interfaces between liquid and air across which there is no mixing but at which the pressure is continuous (because they are vortex sheets, if liquid and air are regarded as parts of a single fluid of piecewise uniform density), and that the liquid jet tends to a uniform stream of width $2kb$, $k < 1$, with increasing distance from the origin. Neglecting gravity and any motion of the air, show how mass conservation, the momentum principle, Postulate VII, and Lemma 12.1 suffice to deduce the limit of k as $b/R \to 0$.

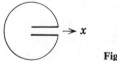

Fig. 11.4.

If the two parallel sheets are absent, all other circumstances being unchanged, is the limit of k greater or less than when the sheets are present?

12. Equations of Motion

By the smoothness convention, and indeed even for those limiting motions for which $\mathbf{v}$ and p_{ij} are of class C^1 on the closure of the fluid domain, the momentum principle can be written as a set of partial differential equations.

The Divergence theorem then implies

$$\int_{\partial\Omega_t} p_{ij}n_j \, dS = \int_{\Omega_t} \frac{\partial p_{ij}}{\partial x_j} \, dV$$

for every fluid domain with regular boundary surface, and since $\rho > 0$ (10.2) is equivalent to

$$(12.1) \qquad Dv_i/Dt = f_i + \rho^{-1} \, \partial p_{ij}/\partial x_j, \qquad i = 1, 2, 3,$$

and this system is commonly called the *equation of motion*.

As noted in Section 10, only motions for which the body force can be absorbed into the pressure will be considered until the end of Chapter 4, and the body force term will accordingly be omitted in the following.

An identity useful in connection with (12.1) is obtained by noting that $v_j(\partial v_j/\partial x_i - \partial v_i/\partial x_j)$ is the i-component of $\mathbf{v} \wedge \text{curl } \mathbf{v}$, and from (3.2) that

$$\frac{Dv_i}{Dt} = \frac{\partial v_i}{\partial t} + v_j \frac{\partial v_i}{\delta x_j} = \frac{\partial v_i}{\partial t} + \frac{\partial(v_j v_j/2)}{\partial x_i} + v_j\left(\frac{\partial v_i}{\partial x_j} - \frac{\partial v_j}{\partial x_i}\right).$$

Since the symbol $|\mathbf{v}|$ is clumsy in many formulae, but v is conveniently reserved for the y-component of $\mathbf{v}$, the symbol q is often used for the velocity magnitude $|\mathbf{v}|$, and then

$$(12.2) \qquad D\mathbf{v}/Dt \equiv \partial\mathbf{v}/\partial t + \tfrac{1}{2} \, \text{grad}\,(q^2) - \mathbf{v} \wedge \text{curl } \mathbf{v}.$$

Any detailed analysis of fluid motion depends on further information on the stress tensor, and the simplest expedient is to neglect the tensor τ_{ij} defined in Section 10 also for nonuniform motion. This is expressed in the

Definition. *Inviscid fluid* means any fluid such that the stress tensor is $p_{ij} \equiv -p(\mathbf{x}, t)\delta_{ij}$.

Experiment indicates, moreover, that this provides a remarkably close approximation to the real stress tensor almost everywhere in the most common motions of the most usual fluids, namely, those "of small viscosity." The concept of inviscid fluid is thus exceedingly useful for the description and analysis of natural and technological fluid motions, and the bulk of the literature on fluid dynamics is based on it.

For the inviscid fluid $\partial p_{ij}/\partial x_j = -\partial p/\partial x_i$, and (12.1) becomes

$$(12.3) \qquad D\mathbf{v}/Dt = -\rho^{-1} \, \text{grad}\, p,$$

which is often called *Euler's equation of motion*. Scalar multiplication by $\mathbf{v}$ and use of (12.2) give

$$(12.4) \qquad \mathbf{v}\cdot\partial\mathbf{v}/\partial t + \tfrac{1}{2}\mathbf{v}\cdot\text{grad}\,(q^2) + \rho^{-1}\mathbf{v}\cdot\text{grad}\, p = 0,$$

which is the general form of *Bernoulli's equation*. Since its first term is $\tfrac{1}{2}\,\partial(q^2)/\partial t$, it is seen to be an energy corollary of the equation of motion.

It is convenient to list here also a number of special cases of (12.3) and (12.4) which are frequently useful. For incompressible motion, where $\rho \equiv$ const, Bernoulli's equation may be written

$$(12.5) \qquad \partial(\tfrac{1}{2}q^2)/\partial t + \mathbf{v}\cdot\text{grad}\left(\frac{p}{\rho} + \frac{1}{2}q^2\right) = 0.$$

If the motion is also irrotational, so that $\mathbf{v} = \text{grad }\phi$, Euler's equation becomes

$$(12.6) \qquad \partial\phi/\partial t + \tfrac{1}{2}q^2 + p/\rho = F(t),$$

by (12.2). The quantity $p + \tfrac{1}{2}\rho q^2 = p_T$ here occurring is called *total pressure* or *stagnation pressure* or *total head* and plays a ubiquitous role in calculations on incompressible motions. For steady flow, finally, (12.5) and (12.6) reduce to Bernoulli's

Lemma 12.1. In steady incompressible flow of inviscid fluid, $p_T = p + \tfrac{1}{2}\rho q^2$ is constant along each streamline, and if the flow is also irrotational, $p_T \equiv$ const on any connected fluid domain.

It is frequently desirable to apply this lemma to limiting motions with discontinuity surfaces at which the assumption $\mathbf{v}, p \in C^1$ of this section fails. The total pressure p_T is then constant across discontinuity surfaces at which only derivatives of $\mathbf{v}$ or p are discontinuous, but not necessarily across discontinuity surfaces at which $\mathbf{v}$ or p itself is discontinuous.

Lemma 12.1 draws attention to the fact that calculations on incompressible motions may easily lead to negative values of p—which appears physically absurd for a quantity meant to represent the thermodynamic pressure. The reason is that the momentum principle (10.1) involves only differences of pressure directly, since $\int \delta_{ij}n_j \, dS = 0$ for any closed, regular surface, and accordingly (12.1) involves only grad p directly. For incompressible motion, in which the density is independent of the pressure, only pressure differences are therefore meaningful, and any convenient reference pressure may be chosen as zero level for p.

Problem 12.1. A fixed cylinder is immersed in steady, incompressible flow of inviscid fluid without body force field. The cross-section of the cylinder is bounded, but of arbitrary shape. Assume that the motion is also irrotational and two-dimensional with velocity normal to the generators of the cylinder. Let x and y denote Cartesian coordinates in the plane of the flow, u and v the corresponding velocity components, and X and Y the corresponding components of the aerodynamic force exerted by the fluid on unit span of the cylinder. Let C denote any simple, counterclockwise oriented circuit in the xy-plane and in the fluid domain such that the cross-section of the cylinder in this plane lies in the subset of the plane *interior* to C. Prove Blasius' theorem that

$$(12.7) \qquad\qquad X - iY = \frac{i\rho}{2} \int_C (u - iv)^2 \, dz,$$

where $i^2 = -1$ and $z = x + iy$. (Problem 10.1 provides a convenient starting point. A useful proof must cover limiting flows in which velocity and pressure fail to have an extension to all of the body surface.)

Problem 12.2. A fluid domain Ω_t is occupied by inviscid fluid in incompressible, irrotational motion under the action of a uniform, constant gravitational field. Let $\Omega(t)$ denote an open bounded subset of Ω_t with regular boundary surface $\partial\Omega$, which depends continuously on t, and let $E(t)$ denote the total energy (kinetic and potential) of the fluid in $\Omega(t)$. Express dE/dt in terms of the values taken on $\partial\Omega$ by the pressure and the derivatives of the velocity potential. Deduce that a subset of $\partial\Omega$ which is a bounding surface of the fluid and on which the pressure is constant contributes nothing to dE/dt.

13. D'Alembert's Theorem

Definition. Ideal fluid means inviscid fluid in incompressible motion.

The usefulness of the inviscid fluid concept has been noted in the preceding section, and that of incompressible motion also provides an excellent description of a large class of fluid motions. Some of the main effects of compressibility will be discussed briefly in the last chapter; for the moment it suffices to note that compressibility is important in the very initiation of fluid motion, in violently unsteady motions, in steady motion at high velocity (roughly, of magnitude greater than about two-thirds of the speed of sound) relative to solid bodies, in motion with a great deal of heat transfer or with chemical reactions, and rarely otherwise. This leaves very many natural and technological fluid motions as a field of application for the ideal fluid concept.

Nonetheless it is clearly a physical abstraction intended to furnish an approximation to the behavior of real fluids, and its fruitful study and interpretation depend on an appreciation of the sense in which it furnishes such an approximation. The understanding of this issue, central to this Introduction, is helped by concentration on a few basic properties of ideal fluids specially relevant to it while by-passing the myriad results on ideal fluid motion found in the literature. It is all the more desirable to stress the existence of this wealth of information beyond the scope of this book, and also its usefulness for those students of fluid dynamics, in particular, who have acquired a grasp of the relation between ideal and real fluid motion. This section will be restricted to certain aspects of steady flows, and we begin by completing the discussion of the cascade (Section 11). It has already been noted that the velocity and the pressure of that flow tend to constants independent of y (Fig. 11.2) as $x \to -\infty$ and also as $x \to +\infty$. The same follows for the total pressure $p_T = p + \rho q^2/2$. Now assume the fluid to be ideal. Then Lemma 12.1 may be applied to the streamline S (Fig. 11.2) to deduce that p_T tends to the same constant as $x \to +\infty$ and $x \to -\infty$, and it follows from (11.7) that $D = 0$ and from (11.8) that

(13.1) $$L = -\rho V\Gamma.$$

As noted in Section 11, this result remains valid for the isolated airfoil, and is then known as Joukowski's theorem. (The negative sign stems from the conventions on the orientation of lift and circulation. More precisely, the theorem applies to an airfoil embedded in steady two-dimensional flow of unbounded ideal fluid with uniform velocity V far upstream. The lift is the force component exerted by the fluid on the airfoil in the direction of increasing y, if that of increasing x coincides with the direction of V; and Γ is the circulation of any counterclockwise oriented circuit in the domain of irrotational motion, which encircles the airfoil precisely once.)

While this result is one of the pillars of aeronautics, it has the disturbing feature of involving the conclusion that the drag D, i.e., the force component in the direction of V, is zero. In steady flight in ideal fluid, lift therefore does not involve drag, i.e., an aircraft can shut down its engines for steady flight. Actually, this conclusion is not valid for three-dimensional motion, but the root of the difficulty is not removed by the more sophisticated picture then emerging. The disturbing lack of drag is a basic feature of ideal fluids, known as d'Alembert's paradox:

Theorem 13.1. Consider a finite body with impermeable surface immersed in a steady stream flowing through a long straight pipe (Fig. 13.1). Specify that the flow tends to a uniform one, far upstream of the body, and assume (as suggested by observation) that it also tends to a uniform one far downstream. Assume the fluid to be ideal. Then the force exerted by the fluid on the body has no component D in the direction of the pipe axis.

Note that this theorem considers three-dimensional flow past a body of arbitrary shape. To prove it, apply the momentum principle to the fluid in a domain Ω bounded by the body surface, the pipe wall, and cross-sections A and B of the pipe located sufficiently far upstream and downstream,

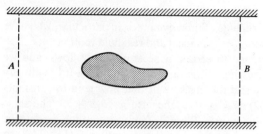

Fig. 13.1.

respectively (Fig. 13.1). Conservation of mass implies that the rates of mass flow across A and B must be equal because the pipe wall and body surface are impermeable. Since the fluid is incompressible and the pipe diameter uniform, it follows that the uniform, axial flows approached far upstream and far downstream must have the same velocity magnitude. The rates at which momentum is transported across A and B, respectively, must therefore differ arbitrarily little, so that the total rate of momentum transport across $\partial\Omega$ is arbitrarily small. The body force field may be ignored if the pressure is understood to be the hydrodynamic one. Since the flow is steady, application of the momentum principle in the form (10.4) therefore shows the resultant force exerted on the fluid in Ω by the stresses on the boundary $\partial\Omega$ to be arbitrarily small.

Now since the velocity field approaches uniformity with distance from the body, it follows from Postulate VII that the same holds for the pressure, whence it follows also for the total pressure, $p_T = p + \rho q^2/2$. By Lemma 12.1, moreover, if a steamline can be traced from A to B, then p_T must approach the same value far downstream as far upstream, and since the same has already been established for the velocity magnitude, it follows also for the pressure. The resultant axial force component exerted across both A and B on the fluid in Ω can therefore be made arbitrarily small in magnitude by choosing those cross-sections sufficiently far from the body. Since the fluid is inviscid, finally, the stress exerted on it across the pipe wall is everywhere normal to the pipe axis. The axial force component, $-D$, exerted by the body on the fluid must therefore be zero.

It remains only to show that a streamline can be traced from A to B (Fig. 13.1).† Now any open subset $S \subset \Omega$ in which $\mathbf{v}$ does not vanish defines a

---

† This is not as obvious at second sight as at first, since the stream may break away from the body surface to cause recirculation regions. Even in their absence it is not easy to argue that any body of fluid which has crossed A (Fig. 13.1) must cross B within a finite time thereafter, because of the occurrence of stagnation points which cannot, by following fluid in its motion, be reached in a finite time from any point of the fluid domain.

stream bundle B_S which is the set of all points of streamlines (Section 4) intersecting S. It is also an open subset of the fluid domain Ω, and its boundary ∂B_S may intersect $\partial\Omega$. The largest subset $\partial B_S'$ of ∂B_S disjoint from $\partial\Omega$ is called a stream tube (this generalizes the definition of Section 4). If $v \neq 0$ at a point of $\partial B_S'$, then, by the definition of the bundle, the streamline through that point must be disjoint from B_S and hence, no mass flow can cross a stream tube. (A mass-flow rate is therefore associated with the stream bundle as an invariant, in the same sense that a vortex flux is associated with a vortex bundle (Section 7).)

In d'Alembert's theorem the velocity is specified to approach a nonzero axial vector far upstream, and the subset S of the cross-section A interior to any circuit C in A therefore also defines a stream bundle B_S. The smoothness convention assures that the tube $\partial B_S'$ is a regular surface. By (4.2), the total mass flow rate $\int \rho v \cdot n \, dS$ across any closed, regular surface vanishes in steady flow, and since $\rho v \cdot n \to$ const $\neq 0$ on A, C cannot be the whole boundary of the tube $\partial B_S'$, i.e., the stream bundle B_S cannot end in the fluid domain. Since the pipe and body surface are impermeable, any such stream bundle B_S must also intersect the cross-section B (Fig. 13.1). Finally, any streamline intersecting A can be defined as the limit of a sequence of such stream bundles, and it then follows that it must intersect B.

It will be observed that the value of the pipe diameter is immaterial, so that d'Alembert's theorem remains valid also for a pipe of diameter arbitrarily large compared with the dimensions of the body. Moreover, reference to a confining wall can be dispensed with altogether when the body and flow have a plane, or axis, of symmetry; the domain Ω can then be defined so that a *net* transport of mass or momentum across its lateral boundary is ruled out on grounds of symmetry. Other extensions of the theorem are indicated by the example of the cascade (in which the uniform streams approached far upstream and downstream are not in the same direction). In all these cases, moreover, a change of Galilean observer suffices to extend the theorem to steady flight of a solid body in a fluid which is at rest, far ahead of the body.

Since the conclusion of d'Alembert's theorem (Theorem 13.1) is clearly at variance with everyday observation, even for fluids of very small viscosity, the theorem has received much scrutiny. But its proof has long come to be regarded as unassailable, and criticism must therefore turn to its assumptions. Here it is to be noted that there can be no direct observation on an ideal fluid and hence the uniformity of the flow far downstream must be in doubt for such fluid. The paradox is the first of a number of difficulties of the theory of inviscid fluid motion to be discussed in the following, which show the approximation of a real, near-inviscid fluid by an inviscid one to be far from straightforward.

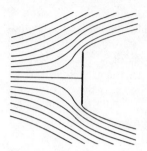

Fig. 13.2. Streamlines of a two-dimensional potential flow past a flat plate to which d'Alembert's paradox does not apply (compare with Figs. 6.5, 6.6, and 6.8).

The reader will have formed a conjecture that the hypotheses of d'Alembert's theorem suffice for deducing the structure of the flow in much more detail. That is commonly done by the help of

Lemma 13.1. If a steady flow of ideal fluid with velocity $\mathbf{v} \in C^2(\Omega)$ is such that there are bounded, open subsets Ω_1 and Ω_2 of Ω with the properties (i) the flow on Ω_1 is arbitrarily close to a uniform, but not stagnant, one and (ii) every streamline intersecting Ω_2 also intersects Ω_1, then the flow on Ω_2 is irrotational.

Proof. The points of Ω_2 can be divided in three classes: viz. (a) points with a neighborhood on which $\mathbf{v} \equiv 0$, (b) points of open sets on which $\mathbf{v}$ does not vanish, and (c) limit points of open sets of points of class (b). By Stokes' theorem (Section 7), it will therefore suffice to show that the vorticity $\omega = 0$ at every point of class (b). Now, the distribution of total pressure on Ω_2 can be deduced from that on Ω_1 by Lemma 12.1. Indeed, since Ω_1 can be chosen so that $|\omega|$ there has an arbitrarily small upper bound, and since (12.2) and (12.3) imply

(13.2) $$\operatorname{grad} p_T = \mathbf{v} \wedge \boldsymbol{\omega}$$

for steady motion of ideal fluid,

$$p_T \equiv \text{const}, \qquad \mathbf{v} \wedge \boldsymbol{\omega} \equiv 0 \quad \text{on } \Omega_2.$$

At any point of class (b), therefore, either $\omega = 0$ or ω is parallel to $\mathbf{v}$, and in any case, $\boldsymbol{\omega} = \lambda \mathbf{v}$ for some scalar field $\lambda(\mathbf{x})$. Since div curl is a null-operator, and since the motion is incompressible, (4.5) implies that λ is constant along each streamline. But on Ω_1, $|\mathbf{v}|$ has a positive lower bound, and $|\lambda|$ therefore has an arbitrarily small upper bound on Ω_2.

Observe that the lemma is also relevant when a solid body moves at a fixed velocity into a large body of fluid which is at rest, far ahead of the body, because such a motion can be steady for an observer moving with the body. This is a very common situation, and it is therefore not surprising that the lemma has played a far-reaching role in theoretical fluid dynamics. It may be noted, however, that little is known yet about streamline patterns of flows with vorticity. The historical role of the lemma has therefore been to identify a priori a large class of ideal fluid motions for which the assumption of irrotational flow can be plausibly consistent and thus to prompt their analysis by the methods of potential theory (Section 5), with results confirming the consistency. Of course, this procedure does not establish uniqueness; in particular, the possibility often remains that there exist also steady, ideal flows with vortex sheets, and even with recirculation regions in which the streamlines are closed (Section 27).

Problem 13.1. Further insight into d'Alembert's paradox may be obtained from consideration of a device designed to produce an axial, aerodynamic force on a body. The simplest mathematical model of a lightly loaded propeller in steady action is a circular disc across which a pressure difference, but no difference in velocity direction, is produced by the action of the blades (the volume of which is neglected). The model takes the fluid to be ideal, and the velocity and pressure to be constant over each side of the disc, and it neglects any velocity component parallel to the disc. The propeller disc moves with constant velocity V in the direction of the axis of symmetry, and the air is taken to be at rest and pressure p_∞, far ahead of the propeller, and it is assumed that the pressure returns to p_∞ far astern, at all distances from the axis.

Show that the model implies a vortex sheet and find the strength of that vortex sheet far astern of the propeller in terms of the thrust coefficient $c_T = T/(\rho V^2 d^2)$, where T is the thrust exerted on the fluid, and d, the propeller diameter. (Since the model turns out to make sense only for a lightly loaded propeller, terms $O(c_T{}^2)$ should be neglected by comparison with unity. The model is, of course, a strongly asymptotic one involving major exceptions to the smoothness convention near the disc.)

14. Kelvin's Theorem

Theorem 14.1. In ideal fluid under a potential body force field, any circuit moving with the fluid conserves its circulation.

Proof. That a circuit C moves with the fluid means (Section 3) that it is the image $H_t C_0$ if a fixed curve C_0 in Lagrangian **a**-space. Its circulation is therefore

$$\Gamma(C) \equiv \int_C \mathbf{v} \cdot d\mathbf{x} = \int_{C_0} v_i(\mathbf{x}(\mathbf{a}, t), t) \frac{\partial x_i}{\partial a_j} \, da_j = G(C_0, t),$$

and the assertion to be proved is that this integral in **a**-space is independent of the parameter t. The argument of Section 3 deducing the smoothness of H_t on $\overline{\Omega_0}$ from the smoothness convention can, by Postulate II, be applied also to H_t^{-1} on Ω_t. If $C \subset \Omega_t$, it is seen in this way that C_0 is also a circuit and that the interchange of limits in

$$\frac{\partial G}{\partial t} = \int_{C_0} \frac{\partial}{\partial t} \left[v_i(\mathbf{x}(\mathbf{a}, t), t) \frac{\partial x_i}{\partial a_j} \right] da_j$$

is justified. Now, $v_i \, \partial^2 x_i / \partial t \, \partial a_j = v_i \, \partial v_i / \partial a_j$, by the definition of the velocity, and since $\mathbf{v}$ is single-valued on Ω_t it follows from Postulate II that it is single-valued also on Ω_0, and

$$\int_{C_0} \frac{\partial}{\partial a_j} |\mathbf{v}|^2 \, da_j = 0,$$

so that

$$\frac{\partial G}{\partial t} = \int_{C_0} \frac{\partial x_i}{\partial a_j} \frac{\partial}{\partial t} v_i(\mathbf{x}(\mathbf{a}, t), t) \, da_j.$$

By the definition of the convective rate of change (Section 3), $\partial v_i(\mathbf{x}(\mathbf{a}, t), t)/\partial t = Dv_i(\mathbf{x}, t)/Dt$, and since the fluid is ideal, Euler's equation of motion (12.3) implies

$$\frac{\partial x_i}{\partial a_j} \frac{Dv_i}{Dt} = \frac{-1}{\rho} \frac{\partial p}{\partial x_i} \frac{\partial x_i}{\partial a_j} = \frac{-1}{\rho} \frac{\partial p}{\partial a_j},$$

the body force field being ignored (Section 10) on the understanding that p is the hydrodynamic pressure. Since p is also single-valued (Postulates VI and VII),

$$\frac{\partial G}{\partial t} = \frac{-1}{\rho} \int_{C_0} \frac{\partial p}{\partial a_j} \, da_j = 0.$$

If $C \subset \overline{\Omega_t}$, it is the limit of a sequence $\{C_n\}$ of circuits in Ω_t, and the theorem follows for the convected circuit C, which is the limit of the sequence of convected circuits coinciding with $\{C_n\}$ at time t.

Lord Kelvin discovered the theorem as the common, Lagrangian source of two theorems which dominate the theory of the ideal fluid. One is the immediate

Corollary 14.1. If a motion of ideal fluid starts from a uniform one at time $t = 0$, then every circuit $C \subset \Omega_t$ has zero circulation at all times.

Consideration of only the reducible circuits similarly yields

Corollary 1.42. If a motion of ideal fluid starts from an irrotational one at time $t = 0$, then it remains irrotational at all times.

These corollaries extend the conclusion of Lemma 13.1 to a much larger class of ideal fluid motions. An example is ocean swell, which may travel from a storm area over thousands of miles to invade initially quiescent bodies of water. On ideal fluid theory, such motions are also tractable by the methods of potential theory.

On the other hand, consider an airfoil starting to move from rest and proceeding gradually to a state of steady flight. Since the motion starts from rest, Corollary 14.1 shows all circuits in the fluid to have zero circulation at all times, and by Joukowski's theorem (Section 13) the airfoil cannot achieve any lift! Fortunately, this conclusion rests on an incorrect interpretation of the deceptively simple Corollary 14.1 (see Appendix 14). Corollary 14.2. similarly may require the strict interpretation of motion noted in Appendix 2 and does not rule out the generation of vortex sheets embedded in the irrotational motion. Moreover, paradoxes analogous to d'Alembert's can be derived from these corollaries if it is overlooked that new questions arise when circuits moving with *near*-ideal fluid are followed over *long* time intervals.

The other decisive theorem derivable from Kelvin's is Helmholtz'

Corollary 14.3. Vortex lines in ideal fluid are convected with the motion of the fluid.

The pervasive influence of this result on natural and technological fluid motions has been noted in Section 8. Its most direct proof is by an analysis due to Cauchy, starting from Euler's equation (12.3) in the form

$$\partial \mathbf{v}/\partial t - \mathbf{v} \wedge \operatorname{curl} \mathbf{v} = -\rho^{-1} \operatorname{grad} p - \tfrac{1}{2} \operatorname{grad} q^2$$

obtained by the help of (12.2). Since $\rho \equiv$ const, the pressure may be eliminated by taking the curl of this equation. Use of the identity

$$\operatorname{curl}(\boldsymbol{\omega} \wedge \mathbf{v}) = (\mathbf{v}\cdot\operatorname{grad})\boldsymbol{\omega} - (\boldsymbol{\omega}\cdot\operatorname{grad})\mathbf{v} + \boldsymbol{\omega} \operatorname{div} \mathbf{v} - \mathbf{v} \operatorname{div} \boldsymbol{\omega}$$

then leads to Helmholtz' equation

(14.1) $$D\boldsymbol{\omega}/Dt = (\boldsymbol{\omega}\cdot\operatorname{grad})\mathbf{v},$$

by (3.1) and (4.5) and because div curl is a null operator. This equation may be integrated as follows. Write $v_i(\mathbf{x}, t) = u_i(\mathbf{a}, \tau)$, and $\omega_i(\mathbf{x}, t) = w_i(\mathbf{a}, \tau)$, where $t = \tau$ and $\mathbf{x} = \mathbf{x}(\mathbf{a}, \tau)$. Then since $u_i = \partial x_i/\partial \tau$ by definition, Helmholtz' equation reads

$$\frac{\partial w_i}{\partial \tau} = w_k \frac{\partial^2 x_i}{\partial \tau \, \partial a_j} \frac{\partial a_j}{\partial x_k}, \qquad i = 1, 2, 3$$

$$= -w_k \frac{\partial x_i}{\partial a_j} \frac{\partial^2 a_j}{\partial \tau \, \partial x_k},$$

because $(\partial x_i/\partial a_j)\, \partial a_j/\partial x_k = \delta_{ik}$. Since also $(\partial a_m/\partial x_i)\, \partial x_i/\partial a_j = \delta_{mj}$, this implies that $w_i \, \partial a_m / \partial x_i$ is independent of τ, and since the initial value of the Jacobian matrix $\partial a_m / \partial x_i$ is unity, Cauchy's equation

(14.2) $$\omega_k(\mathbf{x}(\mathbf{a}, t), t) = \omega_i(\mathbf{a}, 0) \, \partial x_k / \partial a_i$$

follows; it is, of course, essentially equivalent to Kelvin's theorem.

Corollary 14.2 follows immediately. Furthermore, a vortex line in the initial domain Ω_0 is a solution $\mathbf{a} = \mathbf{a}(\sigma)$ of $d\mathbf{a}/d\sigma = \boldsymbol{\omega}(\mathbf{a}(\sigma), 0)$, and by (14.2) that implies

(14.3)
$$\partial x_i(\mathbf{a}(\sigma), t)/\partial \sigma = (\partial x_i/\partial a_j)\omega_j(\mathbf{a}(\sigma), 0)$$
$$= \omega_i(\mathbf{x}(\mathbf{a}(\sigma), t), t).$$

Thus if $L: \mathbf{a} = \mathbf{a}(\sigma)$ is a vortex line in Ω_0, then $H_t L: \mathbf{x} = \mathbf{x}(\mathbf{a}(\sigma), t)$ is a vortex line in Ω_t.

Now, the meaning so far given to the concepts of vortex line and bundle (Section 7) relates to the fluid domain at a fixed time. The conclusion just drawn from (14.3) shows that it is consistent with those kinematical concepts and results to use Corollary 14.3 to define $H_t L$ as "the same" vortex line as L. We thereby extend (in ideal fluid) with respect to time all the remarkable properties of permanence with respect to space of vortex bundles (Section 7).

Equation (14.3) also shows explicitly how the vorticity magnitude in an ideal fluid increases in direct proportion to the length of the local element of vortex line. The relevance of this result to turbulence has already been noted in Section 8. Another implication is the following. There are many instances of motions, such as two-dimensional flow past an airfoil, in which the fluid stream must branch at the nose of a body. If the motion is rotational, the presence of vortex lines must be anticipated, of which, by Corollary 14.3, part is convected past one side of the body while another part is convected past the other side of the body. If a segment of such a vortex line is initially disjoint from the boundary of the fluid domain, it must remain so disjoint, by (2.1); and Corollary 7.5 therefore implies that it must remain continuous. Accordingly, as such a vortex line segment is swept partly past one and partly past the other side of the body, a part of it must in time be lengthened to an arbitrary degree. By (14.3), therefore, the vorticity magnitude cannot remain bounded if the fluid be ideal. In particular, a steady, ideal fluid flow with such vortex lines cannot have a velocity field even of class $C^1(\overline{\Omega_t})$.

The remark should be added that the results of this section depend essentially on the inviscid nature of the ideal fluid, but not on its incompressibility; they are readily extended to fluids (such as a gas of uniform entropy, Section 33) in which the density ρ is a function only of the pressure p [e.g., 14].

Problem 14.1. Use Kelvin's theorem to show that, in *two-dimensional* motion of ideal fluid under a potential body force field, vorticity itself is conserved at any point moving with the fluid. Deduce the same result also from Cauchy's equation (14.2).

Appendix 14

In the application of Kelvin's theorem difficulties may arise which are easily overlooked; conversely, correct application may lead to unexpected conclusions. The two-dimensional motion due to an airfoil starting from rest and accelerating gradually to steady flight furnishes an instructive example. Figures 14.1a and 14.1b indicate the state of the motion at $t = 0$ and $t > 0$, respectively, with C in Fig. 14.1b representing the image $C = H_t C_0$ of the circuit C_0 of Fig. 14.1a.

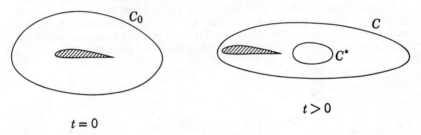

Fig. 14.1a.   **Fig. 14.1b.**

The curve C^* in Fig. 14.1b represents another circuit in the fluid, and to find the corresponding circuit in Ω_0, we must interpret C^* as moving with the fluid and trace its motion back to $t = 0$, just as the motion of C is traced back to C_0 by looking from Fig. 14.1b back to Fig. 14.1a. Comparison of these two figures suggests that the initial position of C^* may well have been as indicated in Fig. 14.1c, so that it did not, at $t = 0$, represent

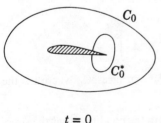

$t = 0$ **Fig. 14.1c.**

a circuit in the fluid domain Ω_0. In that case it follows from Postulate II that the circuit C^* in Fig. 14.1b cannot be a circuit in the fluid domain either! It cannot even be a circuit in the closure $\overline{\Omega_t}$ in the proper sense of being the limit of a sequence of circuits in the interior of Ω_t. It is thus seen that correct application of Corollary 14.1 may depend on the strict interpretation of fluid motion discussed in Appendix 2.

Indeed, the corollary permits us to draw the following, remarkable conclusion of Prandtl, on recalling from Section 3 that the boundary of a body of fluid is a surface moving with the fluid. If an airfoil which started from rest in an ideal fluid is later found to possess nonzero circulation Γ, then the component B of the boundary of the whole body of fluid which coincided with the airfoil at $t = 0$, must coincide at the later time with the union of the airfoil surface and a surface embedded in the fluid which has circulation $-\Gamma$. This follows rigorously from the fact that the circuit C_0 of Fig. 14.1a, and all other circuits in Ω_0 and moving with the fluid, must retain zero circulation, by Corollary 14.1.

The existence of a boundary component embedded in the fluid at $t > 0$ is not, in fact, implausible. To see this, consider the circuit C_0 of Fig. 14.1a as the first of a sequence of disjoint circuits in the fluid, the limit of which is the airfoil surface. A second member, C_0', of such a sequence is indicated in Fig. 14.2a. If the circuits are all regarded as circuits moving with the fluid, and if C_0 is convected to the position C of Figs. 14.1b and 14.2b, then C_0' will be convected to a position such as indicated by the curve C' of Fig. 14.2b. Since the other members of the sequence will be similarly elongated by the convection process, the limit of the sequence in Fig. 14.2b may well be the union of the airfoil contour and a curve such as that indicated by the dashed line of Fig. 14.2b.

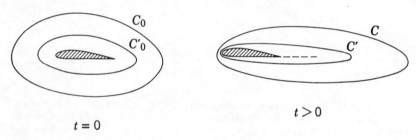

$t = 0$ $t > 0$

Fig. 14.2a.   Fig. 14.2b.

15. Conservation of Angular Momentum

Postulate VIII.   In the absence of a body couple field,

$$\frac{D}{Dt} \int_{\Omega_t} \mathbf{x} \wedge (\rho \mathbf{v}) \, dV = \int_{\Omega_t} \mathbf{x} \wedge (\rho \mathbf{f}) \, dV + \int_{\partial\Omega_t} \mathbf{x} \wedge \mathbf{t} \, dS$$

for every fluid domain Ω_t with regular boundary surface $\partial\Omega_t$, where $\mathbf{t}$ is the vector with components $t_i = p_{ij} n_j$.

Theorem 15.1.   Postulate VIII is equivalent to the condition that the stress tensor is symmetrical, i.e., $p_{ik} = p_{ki}$ for $i, k = 1, 2, 3$.

Proof.   By Corollary 4.1, since $D\mathbf{x}/Dt = \mathbf{v}$ by definition, and by the (linear) momentum principle in the form (12.1),

$$\frac{D}{Dt} \int_{\Omega_t} \rho \mathbf{x} \wedge \mathbf{v} \, dV = \int_{\Omega_t} \rho \mathbf{x} \wedge \frac{D\mathbf{v}}{Dt} \, dV = \int_{\Omega_t} \mathbf{x} \wedge (\rho \mathbf{f} + \mathbf{P}) \, dV,$$

with $P_i = \partial p_{ij}/\partial x_j$. The i-component of $\mathbf{x} \wedge \mathbf{P}$ is

$$\varepsilon_{ijk} x_j \, \partial p_{km}/\partial x_m = \partial(\varepsilon_{ijk} x_j p_{km})/\partial x_m - \varepsilon_{ijk} p_{km} \, \delta_{jm},$$

where ε_{ijk} is the alternating tensor (Section 7), and the Divergence Theorem therefore implies

$$\int_{\Omega_t} \mathbf{x} \wedge \mathbf{P} \, dV = \int_{\partial\Omega_t} \mathbf{x} \wedge \mathbf{t} \, dS - \int_{\Omega_t} \boldsymbol{\pi} \, dV,$$

with $\pi_i = \varepsilon_{ijk} p_{kj}$. Hence Postulate VIII is equivalent to the statement $\pi_i = 0$ for every i, and the theorem follows.

Body couple fields occur in polarized media, and are rare in fluids [but see 15]. In fact, most authors postulate the symmetry of the stress tensor directly. However, Postulate VIII clearly completes a basic set of kinematical and dynamical laws. This set is, broadly speaking, adequate for the description of ideal fluid motion—classically defined by $\tau_{ij} = 0$ and $\rho \equiv \text{const}$—for which Postulate VII already gives an adequate definition of the stress tensor. In this sense Postulate VIII permits us to establish

Theorem 15.2. In the absence of body forces or boundary conditions dependent on the velocity direction, ideal fluid motion on a fixed domain is reversible.

For proof it suffices to verify by inspection that if $\mathbf{v}(\mathbf{x}, t)$ and $p(\mathbf{x}, t)$ satisfy Postulates I to VIII then so do the functions

$$\mathbf{v}'(\mathbf{x}, t) = -\mathbf{v}(\mathbf{x}, -t), \qquad p'(\mathbf{x}, t) = p(\mathbf{x}, -t).$$

An analogous theorem can, of course, be formulated for motion on a time-dependent domain. It should also be noted that the most common boundary conditions are indeed independent of the velocity direction. For instance, the basic kinematic boundary condition (3.4) is so independent, if the boundary surface is fixed (so that $V = 0$) or is permitted to reverse its own motion. Similarly, the physical condition formulating the discontinuity of the pressure at the interface of two fluids in terms of the surface tension does not involve the velocity direction.

It follows immediately that, if air were an ideal fluid, breathing would be possible only in a cross-wind. In its absence, we would always inhale the same body of air which we have just breathed out.

Similarly, water is known to have very small viscosity and to suffer insignificant density changes in ordinary motions, but the theorem shows that ideal fluid theory is inadequate even for explaining why it is impossible to reverse the process of pouring from the spout of a teapot!

Observe, moreover, that the theorem states essentially a property of the *inviscid* fluid, since the proof applies to continuous compressible motion (Section 33) if the density ρ is a function only of the pressure p.

CHAPTER 3

Newtonian Fluid

16. The Couette Experiment

The preceding sections have shown that the concept of ideal fluid, however fruitful it may be in many respects, is yet basically incapable of explaining some of the most commonly observed features of fluid motions. The viscous part of the stress, even if small, must have a decisive influence, and the first question must be how this stress is related to other observable properties of common fluid motions. The primary source of information must be experiment, for we could propose many such relations (called constitutive equations) but would soon discard those not consistent with observation.

Fortunately, a single experiment can furnish the most important information if the momentum principle is used to interpret the observations. Couette's experiment uses two long, straight, concentric circular cylinders; the space between them is filled with fluid. Initially the cylinders and fluid are at rest, and the inner cylinder is then made to rotate slowly. The outer cylinder is left free to rotate or not. It is observed that the fluid begins to move and the outer cylinder also begins to rotate in the same sense as the inner.

The velocity of the inner boundary is therefore seen to be transmitted to the fluid, and this cannot be due to a normal stress, since the boundary moves only tangentially (i.e., in the direction normal to the axis of rotation and tangential to a circular cylinder concentric with that axis). The observed motion of the outer cylinder, moreover, permits the inference, in view of the momentum principle, that a tangential stress is transmitted through the fluid to the outer boundary.

A further observation may be made if the inner cylinder is driven so that its rate of rotation becomes constant eventually. The fluid and outer cylinder are then ultimately seen to rotate with the same angular velocity as the inner cylinder. Since no tangential acceleration remains, it may be inferred that there is no tangential stress when the fluid rotates like a solid body.

These two qualitative inferences may be used to formulate a mathematical expression for the viscous stress by means of the following classical argument.

17. Constitutive Equation

From Postulate VII (Section 10), the pressure p is known to have the property that $p_{ij} + p\,\delta_{ij} \equiv \tau_{ij}$ vanishes in uniform motion, i.e., when $\partial v_k/\partial x_m = 0$ for $k, m = 1, 2, 3$, while the velocity $\mathbf{v}$ itself is arbitrary. Now $\partial v_k/\partial x_m$ can be split into its symmetrical part e_{km} and antisymmetrical part α_{km} (Section 7), and Problem 7.1 shows that $e_{km} \equiv 0$ in solid body rotation, while the components of α_{km} are then constants proportional to the angular velocity. The second observation of the Couette experiment (Section 16) therefore shows that there is no shear stress, i.e., the off-diagonal components of p_{ij} vanish, when $e_{km} \equiv 0$ and the components of α_{km} are arbitrary constants. The first observation of the experiment, on the other hand, shows that a shear stress can be transmitted under circumstances in which $\partial v_k/\partial x_m \neq 0$. Accordingly τ_{ij} may be expected to depend on e_{km}, and the simplest possibility is

$$(17.1) \qquad \tau_{ij} = 2\mu e_{ij},$$

where the *viscosity* μ is a scalar quantity independent of the velocity and its derivatives. (Usually μ depends only on temperature and chemical composition.) The meaning of "simplest" is clarified by

Theorem 17.1 If p_{ij} is symmetrical, linear in $\partial v_k/\partial x_m$, independent of $\mathbf{v}$, $\partial \mathbf{v}/\partial t$, α_{km}, and higher derivatives of $\mathbf{v}$, and if the relation between the tensors p_{ij} and e_{km} is isotropic (i.e., invariant under rotation of the coordinate system), then $p_{ij} = -p\,\delta_{ij} + 2\mu e_{ij} + \mu'\,\delta_{ij}e_{kk}$, where p, μ, and μ' are scalars.

Proofs will be found in [16] and other books on continuum mechanics or tensor analysis. We therefore adopt

Postulate IXa. $p_{ij} = -(p - \mu'e_{kk})\,\delta_{ij} + 2\mu e_{ij}.$

This defines the Newtonian fluid, in contrast to less common types of liquids to which more complicated constitutive equations apply. Since the trace $e_{kk} = \operatorname{div}\mathbf{v}$, (4.5) shows the postulate to agree with (17.1) for incompressible motion of Newtonian fluid, to which the considerations of Chapters 3 to 5 will be restricted.

A simple example will be useful. Consider fluid between two parallel, infinite, plane solid surfaces, one of which is at rest while the other moves with a constant velocity in its own plane. The direction of this velocity will be taken as that of increasing x, and the y-axis, as normal to the planes (Fig. 17.1). Assume that the x-component of the fluid velocity depends only on y,

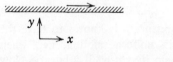

Fig. 17.1.

i.e., $v_1 = u(y)$, and $v_2 = v_3 = 0$; this is an obvious approximation to the situation encountered in the Couette experiment when the gap between the two cylinders is small compared with the radius of either of them. It follows that $e_{ij} = \frac{1}{2}(\partial v_i/\partial x_j + \partial v_j/\partial x_i) = 0$ unless $i, j = 1, 2$ or $2, 1$, and $e_{12} = e_{21} = \frac{1}{2}\, du/dy$. According to (17.1), therefore, the only nonzero components of τ_{ij} are

$$(17.2) \qquad\qquad \tau_{12} = \tau_{21} = \mu\, du/dy,$$

as was suggested implicitly by the first observation in Section 16.

It should be mentioned that air, water, and most common industrial fluids are Newtonian under virtually all conditions, but many plastic liquids are not, having viscous stresses τ_{ij} dependent on α_{km}, on anisotropies of their internal structure, on higher-velocity derivatives, or even on the history of their motion. Many more experiments are then needed to infer the form of the constitutive equations, and the study of these questions, called rheology, is advancing rapidly.

The approach of continuum theory to constitutive equations, of which the reasoning of this section is a classical example, has the important advantage of requiring no distinction between gases and liquids. A disadvantage, on the other hand, is that this type of argument yields no information on how the viscosity could be calculated or even what its value might depend on. A much more crucial drawback, however, is the following. Substitution of (17.1) into (12.1) shows that the equation of motion of a Newtonian fluid must involve second derivatives of the velocity, in contrast to (12.3), which involves only the first derivatives. It is mathematically plausible, therefore—and amply confirmed below—that the Newtonian fluid must require an additional boundary condition beyond that appropriate to the ideal fluid. But in its present state, continuum theory provides no direct clue to that boundary condition.

It is fortunate, therefore, that a complementary approach is offered by kinetic theory. This has not yet been developed to a satisfactory stage for liquids. But for gases, it is capable of predicting not only the constitutive equations, but also the corresponding boundary condition, and moreover, it shows that those two form a single conceptual unit.

18. Kinetic Theory

Since only the very simplest, qualitative results are needed here, the following is restricted to the roughest sketch of the most elementary part of the theory. Loose ends will be left at every step, but it will become clear that these can have no appreciable effect on the main conclusions that will be drawn. It is indeed a triumph of kinetic theory that so superficial an argument already leads convincingly to such penetrating conclusions. For better accounts the reader is referred, e.g., to [17] and thence to [18] and [19].

Kinetic theory pictures the simplest type of gas, say hydrogen, as a collection of molecules, each of the same mass m. The number n of molecules per unit volume is very large, but the volume actually occupied by any one molecule is extremely small compared with the nth fraction of the unit volume (which is understood to be the unit volume of a typical fluid dynamical experiment, say 1 cm^3). In fact, each molecule spends most of its life in free flight, out of the range of influence of any other molecule, but the number n is so large that each suffers many collisions with others in the unit time of the experimenter (or "mascroscopic observer"). The average—over all the molecules in a unit volume—distance traveled by a molecule in free flight between successive collisions is called mean free path l, and is small compared with the macroscopic unit of length. Relations between macroscopic properties of a gas may then by inferred by considering the flight and collisions of molecules from the point of view of particle mechanics, deducing relations between averages of mechanical quantities, and then evaluating those relations in a limit $l \to 0$.

To discuss some aspects of momentum transport in this manner, let $\mathbf{w}$ denote the velocity of an individual molecule. The average of $\mathbf{w}$ over many molecules, or "mean velocity," will be called $\mathbf{v}$, since that is the kinetic interpretation of the fluid velocity defined by Postulate IV (Section 2). Thus $\mathbf{w} = \mathbf{v} + \mathbf{c}$, where, by definition, the mean of $\mathbf{c}$ is zero, so that the simplest measure of the random part $\mathbf{c}$ of the velocity is the root mean square of $|\mathbf{c}|$, denoted by c. The limit envisaged here is one in which cl tends to a nonzero limit (in the macroscopic units) as $l \to 0$. This will be seen in Sections 23 and 33 to cover the case of incompressible motion in particular.

Consider now a flat geometrical area A in the xz-plane (Fig. 18.1), in a stream of molecules of mean velocity $\mathbf{v} = \{u(y), 0, 0\}$; this will give the kinetic view of the idealized Couette experiment (Section 17). The area A, however, is not at first to be identified with a solid wall; it is freely penetrable by molecules, representing simply the place where they will be counted.

It is thoroughly plausible that the number N of molecules reaching A from below ($y < 0$) in a short time interval (t_1, t_2) is proportional to n, to A, and to $t_2 - t_1$. Since A is parallel to the particular stream of molecules, N cannot

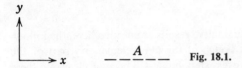

Fig. 18.1.

depend on $\mathbf{v}$ but should plausibly depend on c. Hence $N = kcnA(t_2 - t_1)$, where k is a nondimensional quantity; it will not be necessary to inquire what it depends on. These N molecules must have had their last collision before arriving at A, on average, at $y = -k_1 l$, with $0 < k_1 < 1$. Their mean velocity will be determined mainly by their last collision, and for simplicity it will be assumed to be entirely so determined, so that they have a mean velocity appropriate to the place of their last collision, i.e., $\mathbf{v} = \{u_1, 0, 0\}$, $u_1 = u(-k_2 l)$, $0 < k_2 < 1$. Assume that the process is steady (in the mean), then the same number N must reach A from above in the time interval (t_1, t_2), and those have mean velocity $u(k_2 l) = u_2$.

The net flow rate upward through A of x-component of momentum is therefore

$$F = Nm(u_1 - u_2)/(t_2 - t_1) = kcnmA(u_1 - u_2),$$

and since $nm = \rho$ and $(u_2 - u_1)/(2k_2 l) \to du/dy$ as $l \to 0$ (with du/dy taken at $y = 0$),

$$F \to -k_3 c\rho l A \, du/dy,$$

with $k_3 > 0$ and nondimensional, and with cl standing for the limit of this product. By the momentum principle (Section 10), F must be the x-component of the force exerted across A by the fluid below A on the fluid above. The natural sense of the positive normal on A is that of y increasing, and since $p_{ij}n_j$ is defined (Section 10) as the force exerted by the outside fluid (into which $\mathbf{n}$ points) on the inside fluid, $F = -p_{12}A$, whence

(18.1) $$p_{12} = \tau_{12} \to k_3 \rho cl \, du/dy.$$

This confirms (17.2) and shows that

(18.2) $$\mu = k_3 \rho cl.$$

Of course, the argument can be readily generalized to give $\tau_{ij} = 2\mu e_{ij} + \mu' e_{kk} \, \delta_{ij}$ for all i, j, and a closer look at the kinetic model shows k_3 to be a number dependent on the nature of the collisions [17]; for instance, for a monatomic gas with spherical, elastic, smooth atoms, $k_3 = (5/16)(6\pi)^{1/2}$.

In fluid dynamics, the *kinematic viscosity* $\nu \equiv \mu/\rho$ is usually a more convenient quantity than μ. From (18.2)

(18.3) $$\nu = k_3 cl.$$

Although it has long been known to some physicists, the fluid dynamical literature has but recently acknowledged that the kinetic explanation of viscous shear must have immediate implications for the behavior of a gas near a solid surface. Consider the special case of an area A (Fig. 18.1) which is part of a plane solid boundary, $y = 0$, in the frame of an observer for whom the solid wall is at rest and the gas occupies the half-plane $y > 0$. Since the process is assumed steady, the same number of molecules reach and leave A in any given time interval. The incident molecules again have mean velocity $\mathbf{v} = \{u_2, 0, 0\}$ with $u_2 = u(k_2 l)$.

There are two basic, extreme reflection processes. One, called specular reflection, is defined so that the molecules leaving A also have mean velocity u_2, and in that case the limit $u(+0)$ of the mean velocity, as $y \downarrow 0$, is also u_2. The other, called diffuse reflection, is defined so that the molecules leaving A have zero mean velocity, because they are captured by the electrostatic field, and later kicked off randomly by the thermal oscillations, of the solid crystal. In that case the limit is $u(+0) = u_2/2$.

Real reflection is always a mixture of the two, so that the molecules leave A with mean velocity $(1 - \alpha)u_2, 0 < \alpha < 1$. Unless special precautions are taken, the diffuse reflections are found to predominate, even on smooth surfaces, and practical surfaces are usually very rough on the scale of the mean free path, so that $1 - \alpha \ll \alpha$. In any case the limit of the mean velocity must reflect the mixture, so that the *slip velocity* $\lim_{y \downarrow 0} u(y) = u(+0)$ is

$$u(+0) = \tfrac{1}{2}u_2 + \tfrac{1}{2}(1 - \alpha)u_2$$

with $u_2 = u(k_2 l)$. But also

$$\frac{u(k_2 l) - u(+0)}{k_2 l} \to \lim_{y \downarrow 0} \frac{du}{dy} = \left(\frac{du}{dy}\right)_{+0}$$

as $l \to 0$, so that u_2 may be eliminated to obtain $u(+0)/l \sim k_4 (du/dy)_{+0}$, and from (18.1)

$$(18.4) \qquad \frac{u(+0)}{\tau_{12}(+0)} \to \lim \frac{lk_4}{\mu} = \lim \frac{2 - \alpha}{2\alpha k \rho c} = 0$$

as $l \to 0$ for cl fixed. But the integral over $\tau_{12}(+0)$ is the force on the solid boundary, which must be finite for finite solid surface area. Hence the slip velocity $u(+0)$ must tend to zero on the solid surface, and we must adopt the final

Postulate IXb.   The boundary condition at a solid surface is zero relative fluid velocity,

both with respect to the tangential and normal velocity components (in view of (3.4)).

Of course, solid boundaries are not the only boundaries of fluids. But a discussion of all boundary conditions relevant to Newtonian fluids would lead too far afield, and we shall be content here with Postulate IXb as a representative sample.

It may be recalled that the inviscid fluid is classically defined (Section 12) by $\tau_{ij} = 0$ for all i, j. The result (18.4) thus implies Postulate IXb for the inviscid fluid under an even larger class of kinetic limits. In other words, the kinetic model implies inescapably that the *inviscid* fluid can have strictly *no slip* at a solid surface!

This will appear to be quite an absurd conclusion to most readers, since it contradicts the standard statement that an inviscid fluid is free to slip on a solid surface. It also appears inconsistent (Section 19) with the mathematical requirement of well-posure. But a second reading of this section will show that rejection of the "absurd" conclusion would imply rejection of the basic kinetic model. The present conclusion is, in fact, correct without the slightest doubt, and this was known to Prandtl already in 1904. The apparent contradiction in it is only yet another form of the ideal fluid paradox, which will be elucidated in Chapter 4 by the help of Postulate IXa, b.

Appendix 18

Heat conduction. It is worth noting for reference in Chapter 6 that a quite analogous argument concerning the transport of energy, rather than momentum, across A gives the kinetic explanation of heat conduction in a gas [17]. In the same limit $l \to 0$, the random part of the molecular motion is found to define a *heat flux* vector field $\mathbf{q}$ such that $\mathbf{q} \cdot \mathbf{n}$ is the rate at which heat is conducted across unit area of a surface in the sense of the unit normal $\mathbf{n}$. Moreover, this flux is related to the temperature T by

$$(18.5) \qquad\qquad \mathbf{q} = -\lambda \operatorname{grad} T.$$

The same result, of course, may be obtained by general considerations, analogous to those of Theorem 17.1, concerning the simplest relation between heat flux and temperature in any continuous medium, and (18.5) was first proposed by Fourier in this manner. In effect, it is most logical to regard this relation simply as a part of Postulate IXa.

The distinctive contribution of kinetic theory to Fourier's law (18.5) is to give an explanation of both the conduction of heat and the transmission of shear stress (and diffusion, too) in gases in terms of the same process of molecular flights and collisions. That implies a relationship between the *coefficient* λ *of heat conduction* in (18.5) and the viscosity in (17.2). It is that, with c_p denoting the specific heat at constant pressure, the nondimensional Prandtl number

$$\sigma = \mu c_p / \lambda$$

is found to be virtually constant in a gas (its value is usually near 0.7).

It follows that an inviscid gas should logically also be considered incapable of conducting heat. Conversely, if viscous effects are taken into account, it is inconsistent to neglect heat conduction if appreciable temperature differences are present.

19. Some Viscous Fluid Motions

With the constitutive equation of Postulate IXa, the equation of motion (12.1) becomes

$$(19.1) \qquad \rho \frac{Dv_i}{Dt} = -\frac{\partial}{\partial x_i}\left(p - \mu' \frac{\partial v_k}{\partial x_k}\right) + \frac{\partial}{\partial x_j}\left[\mu\left(\frac{\partial v_i}{\partial x_j} + \frac{\partial v_j}{\partial x_i}\right)\right].$$

If no substantial temperature gradients occur, it is usually sufficient to approximate μ and μ' by constants, and for incompressible motion (19.1) then reduces to

$$(19.2) \qquad Dv_i/Dt = -\rho^{-1}\,\partial p/\partial x_i + \nu\,\nabla^2 v_i, \qquad i = 1, 2, 3$$

by (4.5). In vector notation this is

$$(19.3) \qquad D\mathbf{v}/Dt = -\rho^{-1}\,\mathrm{grad}\,p + \nu\,\nabla^2\mathbf{v},$$

where ∇^2 denotes the Laplacian operator $\partial^2/\partial x_k\,\partial x_k$. This approximate form of the full equation of motion will be taken as the basis of the discussion in this section and Chapters 4 and 5, not only because it furnishes a good approximation for many important motions, but also because it is adequate for explaining the key features of the relation between viscous and ideal fluid motions.

It is of interest to note the forms which some of the salient relations of ideal fluid dynamics take for this incompressible viscous fluid. Bernoulli's equation (12.4) for steady flow was obtained from (12.3) by multiplying both sides scalarly by $\mathbf{v}$ and using (12.2); applied to (19.3), that operation yields

$$\mathbf{v}\cdot\mathrm{grad}\,p_T = \nu\mathbf{v}\cdot(\nabla^2\mathbf{v})$$

for steady flow. Viscosity is thus seen to cause a variation of total pressure along streamlines, and Lemma 13.1 and d'Alembert's Theorem 13.1 do not apply to a viscous fluid. Similarly, (19.3) is not invariant under the transformation of the reversibility theorem 15.2. Use of (19.3) in place of (12.3) in the proof of Kelvin's Theorem 14.1 leads to

$$D\Gamma/Dt = \nu \int_G (\nabla^2\mathbf{v})\cdot d\mathbf{x},$$

so that viscosity is also seen to cause a variation in the circulation of all but some very exceptional circuits moving with the fluid. The viscous term takes a more illuminating form in Helmholtz' equation (14.1), which becomes

$$(19.4) \qquad D\boldsymbol{\omega}/Dt - (\boldsymbol{\omega}\cdot\mathrm{grad})\mathbf{v} = \nu\,\nabla^2\boldsymbol{\omega}$$

if the derivation of Section 14 is based on (19.3) rather than (12.3). Equation (19.4) has a mathematical structure similar to that of the standard heat conduction, or diffusion, equation $\partial\phi/\partial t = \nabla^2\phi$. Thus the right-hand term of (19.4) may be interpreted as representing viscous diffusion of vorticity—much like diffusion of temperature in heat conduction—while the left-hand side has been shown in Section 14 to represent convection of vortex lines with the fluid.

One of the simplest, practically important examples of viscous fluid motion is steady flow in an infinitely long, straight, circular pipe. We take cylindrical polar coordinates x, r, θ and denote the corresponding velocity components by u, v, w, respectively. In view of the symmetry of the problem, existence of a flow may be anticipated for which the velocity field is independent of the axial coordinate x and is axially symmetrical and without swirl (Section 4), and attention will be restricted here to such flows. Then $w \equiv 0$, and u and v depend only on r. By (4.8), $rv(r) = \text{const}$, and since the pipe wall is taken to be an impermeable, solid surface, Postulate IXb implies

(19.5) $$v \equiv 0.$$

It follows from the equations (19.3) for the radial and circumferential velocity components that the pressure p can depend only on x. The velocity field $u(r)$ and pressure field $p(x)$ are therefore governed by the remaining equation of motion

$$Du/Dt = -\rho^{-1}\, dp/dx + v\, \nabla^2 u,$$

with the boundary condition

(19.6) $$u(a) = 0$$

implied by Postulate IXb, where a denotes the pipe radius, and the regularity condition,

(19.7) $$u(0) \text{ exists,}$$

implied by Postulate IV. Problem 4.2 gives the divergence operator for a vector field dependent only on r, and since the gradient operator for such a function has the components $0, d/dr, 0$, (5.3) shows $\nabla^2 u$ to reduce to $r^{-1} d(r\, du/dr)/dr$ in the case under discussion. By (19.5), moreover, Du/Dt is seen to vanish, so that the differential equation for u and p reduces to

(19.8) $$\frac{dp}{dx} = \frac{\mu}{r}\frac{d}{dr}\left(r\frac{du}{dr}\right).$$

It is instructive to consider first ideal fluid, i.e., the case $\mu = 0$. Then (19.8) implies $p = \text{const}$, and $u(r)$ is indeterminate? Indeed, the influence of conditions at the pipe "inlet" $x = -\infty$ persists undiminished for all x in this case. The inlet condition of most practical interest is that the flow there is near-uniform and, in any case, irrotational. Lemma 13.1 then shows the flow to remain irrotational for all x, and since the cylindrical components of the vorticity curl $\mathbf{v}$ are $0, 0, -du/dr$ when $\mathbf{v}$ has components $u(r), 0, 0$, (19.6) implies that there is *no* nontrivial solution! (If the irrotational condition is abandoned, on the other hand, any velocity profile $u(r)$ subject to (19.6) and

(19.7) may be specified, and any mass flow of ideal fluid can be driven through any length of pipe without pressure loss!)

These difficulties are not peculiar to pipe flow, but are symptoms of a general, apparent incompatibility between Postulate IXb and the ideal fluid concept. That incompatibility is indicated by Corollary 6.2 and even more starkly by

Jeffreys' paradox [20]. Consider the motion of ideal fluid bounded by a tank and a finite number of immersed bodies, all of which are undeformable and have impermeable, regular boundary surfaces. Then if the fluid domain Ω_t is bounded and simply connected, and if the velocity is continuous on $\bar{\Omega}_t$, no relative motion of fluid and bodies exists which started from rest.

The fluid and bodies are therefore capable only of a rigid translation; none of the bodies can even be made to rotate! For a proof, see Appendix 19.

The situation is quite different for a viscous fluid. Since $p = p(x)$, but $u = u(r)$, (19.8) implies

$$dp/dx = \text{const} = \mu A,$$

say, and $u = -Ar^2/4 + c_1 \log r + c_2$, and (19.7) and (19.6) show the constants of integration to be $c_1 = 0$ and $c_2 = Aa^2/4$, so that Poiseuille's solution,

$$u(r) = \frac{r^2 - a^2}{4\mu} \frac{dp}{dx} = \frac{2Q}{\pi \rho a^2} \left(1 - \frac{r^2}{a^2}\right),$$

is obtained, where $Q = 2\pi \rho \int_0^a r u(r) \, dr$ is the total mass flow rate. There is therefore a unique steady flow independent of x and θ and of inlet conditions, which may be approached with increasing distance from the pipe inlet.

It is worth noting that Poiseuille's flow is an exact solution of the nonlinear partial differential system (4.5), (19.3), sometimes referred to as Navier-Stokes equations (the full system of those equations will be found in Section 38). The number of known exact solutions is not very large, although the term "exact" is conventionally extended to cover not only explicit solutions, like Poiseuille's and those of Problems 19.1 to 19.3, but also any solution obtainable by (numerical) integration of a system of ordinary differential equations. The exact solutions are therefore particular solutions distinguished by symmetries of one kind or another; surveys of such solutions are found in [21] and [22].

Problem 19.1 (Diffusion of a Vortex Sheet). An incompressible Newtonian fluid under a potential body force field fills all of space. At time $t = 0$ the fluid in $y > 0$ is in uniform motion parallel to the plane $y = 0$, and that in $y < 0$ is also in uniform motion, but with velocity equal and opposite to that of the fluid in $y > 0$. Find a motion of the fluid for $t > 0$ consistent with this initial condition and such that the velocity tends to the respective initial values as $y \to \pm\infty$ for finite t. Calculate the vorticity field and note its analogy with the temperature field of a simple heat conduction problem. Deduce a solution of "Rayleigh's Problem" on diffusion of vorticity from a solid boundary, which is formulated as follows. The fluid fills the half-space $y > 0$ above the solid, impermeable wall $y = 0$. Fluid and wall are at rest at $t = 0$, but for $t > 0$ the solid wall moves with a fixed velocity in its own plane.

Problem 19.2. The half-space $y > 0$ above the solid, impermeable plane $y = 0$ is filled with incompressible, Newtonian fluid under a potential body force field. The solid wall performs a linear oscillation in its own plane with velocity $U \cos nt$ parallel to a fixed direction, where t is the time and U and n are constants. Find a motion of the fluid consistent with these conditions which is periodic in t, with phase linear in y, and such that the velocity tends to zero as $y \to \infty$. How does the thickness of the moving layer depend on the period and the viscosity?

Problem 19.3 (Asymptotic Suction Profile). The half-space $y > 0$ above the plane solid wall $y = 0$ is filled with incompressible Newtonian fluid in steady flow under a potential body force field. The wall $y = 0$ is porous, and fluid is sucked through it in the normal direction so that the y-component of velocity at $y = 0$ is a constant. Find a motion such that the velocity tends to a uniform one as $y \to \infty$.

Problem 19.4 (Two-dimensional Stagnation Point Flow). The half-space $y > 0$ above the solid impermeable plane $y = 0$ is filled with incompressible Newtonian fluid under a potential body force field. Prandtl conjectured that a two-dimensional steady flow exists such that $u = \beta x F(y)$, $v = G(y)$, with $\beta = $ const. Show that this requires a particular dependence of the pressure on v. Find the differential equation for G (called Hiemenz' equation; it is of third order) and the boundary conditions for it which ensure that the total pressure tends to a limit independent of x as $y \to \infty$. (A proof of existence and uniqueness of the problem thus formulated is given in [23].) Find the potential $\phi(x, y)$ of the motion to which the flow tends as $y \to \infty$, neglecting a function only of y which is $o(y^2)$, and compare the equipotential lines with those of Fig. 6.1. Show that the motion tends to this potential flow already as $y(\beta/\nu)^{1/2} \to \infty$.

Appendix 19

Jeffreys' paradox [20] may be proved as follows. Since the fluid is ideal and the motion starts from rest, Corollary 14.2 shows the motion to remain irrotational, and the continuity of the velocity excludes vortex sheets. Since Ω_t is simply connected, the motion thus has a single-valued potential $\phi(\mathbf{x}, t)$ on $\overline{\Omega}_t$.

Denote by $\mathbf{r}_k$ the position vector with respect to the centroid of the kth body, by $\mathbf{V}_k$ the velocity of that centroid, and by $\boldsymbol{\alpha}_k$ the angular velocity of that body. Then rigid body kinematics shows the velocity of any point of the kth body to be $\mathbf{v}_k = \mathbf{V}_k + \boldsymbol{\alpha}_k \wedge \mathbf{r}_k$. Since the fluid velocity is continous on $\bar{\Omega}_t$, Postulate IXb implies that the circulation of any circuit in the boundary ∂B_k of the kth body is $\int \mathbf{v}_k \cdot d\mathbf{x}$, and this must vanish because the potential is single-valued (Section 5). But since $\boldsymbol{\alpha}_k \wedge \mathbf{r}_k$ (and hence also $\mathbf{v}_k$) is continuously differentiable on a neighborhood of that boundary ∂B_k, which is a regular surface, it follows from Stokes' theorem (Section 7) that $\mathbf{n} \cdot \mathrm{curl}\ \mathbf{v}_k \equiv 0$ on ∂B_k if $\mathbf{n}$ denotes the unit normal on it.

In terms of the alternating tensor ε_{ijl} (Section 7) and of the respective components α_m and r_n of $\boldsymbol{\alpha}_k$ and $\mathbf{r}_k$,

$$\mathbf{n} \cdot \mathrm{curl}\ \mathbf{v}_k = n_i \varepsilon_{ijl}\ \partial(\varepsilon_{lmn}\alpha_m r_n)/\partial x_j,$$
$$= n_i \varepsilon_{ijl} \varepsilon_{lmn} \alpha_m\ \delta_{nj},$$

and it is readily verified that $\varepsilon_{ijl}\varepsilon_{lmj} = \delta_{im}$. Thus

$$\mathbf{n} \cdot \mathrm{curl}\ \mathbf{v}_k = n_i \alpha_i = \mathbf{n} \cdot \boldsymbol{\alpha}_k$$

is the quantity vanishing everywhere on the surface of the kth body B_k (inclusive of the tank, which will be regarded as B_0). If $\boldsymbol{\alpha}_k \neq 0$, therefore, ∂B_k must be an infinite cylinder contrary to the hypothesis that Ω_t is bounded. Hence $\boldsymbol{\alpha}_k = 0$ for all k, i.e., none of the solids rotates.

This implies that the potential ϕ has the boundary values $\phi_k = \mathbf{V}_k \cdot \mathbf{r}_k + c_k$ on ∂B_k, where c_k is a scalar constant, and since this function satisfies Laplace's equation, it extends the potential throughout the interior of the kth body B_k with continuous gradient. Choose any $\mathbf{x} \in \Omega_t$ and let r denote distance from $\mathbf{x}$. Then if $D \subset E^3$ is open and $\mathbf{x} \notin D$, it is readily verified that $\nabla^2(1/r) = 0$ on D. Let Σ denote a sphere about $\mathbf{x}$ large enough to enclose the tank surface ∂B_0 and choose D as the domain between Σ and a sufficiently small sphere about $\mathbf{x}$. Then the potential ϕ on Ω_t with the indicated extension to $\bar{D}$ satisfies $\nabla^2 \phi = 0$ on D and has the boundary values $\phi_0 = \mathbf{V}_0 \cdot \mathbf{r}_0 + c_0$ on Σ.

Now $\phi \nabla^2 \psi = \mathrm{div}\ (\phi\ \mathrm{grad}\ \psi) - (\mathrm{grad}\ \psi) \cdot (\mathrm{grad}\ \phi)$, and since $\phi \in C^1(\bar{D})$ and ∂D is a regular surface the Divergence theorem implies

$$0 = \int_D \left(\frac{1}{r}\ \nabla^2 \phi - \phi\ \nabla^2\ \frac{1}{r} \right) dV$$
$$= \int_{\partial D} \left(\frac{1}{r}\ \mathbf{n} \cdot \mathrm{grad}\ \phi - \phi \mathbf{n} \cdot \mathrm{grad}\ \frac{1}{r} \right) dS.$$

The small sphere may now be shrunk to the point $\mathbf{x}$ to obtain

$$4\pi\phi(\mathbf{x}, t) = \int_\Sigma \left(\frac{1}{r}\ \mathbf{n} \cdot \mathrm{grad}\ \phi - \phi \mathbf{n} \cdot \mathrm{grad}\ \frac{1}{r} \right) dS,$$

and since r and grad ϕ are constants on Σ, $\int_\Sigma r^{-1} \mathbf{n} \cdot \mathrm{grad}\ \phi\ dS = 0$. Thus

$$\phi(\mathbf{x}, t) = (4\pi r^2)^{-1} \int_\Sigma \phi\ dS = c_0 + \mathbf{V}_0 \cdot \langle \mathbf{r}_0 \rangle,$$

where $\langle \mathbf{r}_0 \rangle$ denotes the mean of $\mathbf{r}_0$ over Σ, i.e., the position vector of the center $\mathbf{x}$ of Σ with respect to the centroid of the tank. Hence grad ϕ is constant throughout Ω_t.

The resolution of the paradox lies in the observation that the theorem need not apply to the ideal limit of a Newtonian fluid motion, because the assumption $\mathbf{v} \in C(\bar{\Omega}_t)$ need not apply to such a limit motion. Even if it be relaxed only to $\mathbf{v} \in C(\Omega_t)$, this velocity field of the limit motion need no longer satisfy a physical boundary condition on a solid surface. Indeed, the paradox was proposed by Jeffreys [20] to emphasize the nontrivial nature of the ideal limit of Newtonian fluid motion.

CHAPTER 4

Fluids of Small Viscosity

20. Reynolds Number

The practical usefulness of the inviscid fluid concept (Section 12) derives from the fact that the most commonly encountered fluids have small viscosity. Their behavior should therefore be approximable by that of inviscid fluids. However, the paradoxes of Sections 13, 15, and 19 will have made plain that the successful approximation of real fluid motion by inviscid fluid motion cannot be at all straightforward. This chapter is therefore devoted to a more direct discussion of real fluid motion past solid bodies *in the limit of zero viscosity*.

To this end, we must first of all inquire what real, nondimensional meaning may be attached to the phrase "small viscosity." The equations governing the motion of incompressible Newtonian fluid are

(4.5) $$\operatorname{div} \mathbf{v} = 0$$

(19.3) $$D\mathbf{v}/Dt = -\rho^{-1} \operatorname{grad} p + \nu \, \nabla^2 \mathbf{v}.$$

They contain only two dimensional constants, ρ and ν, and to make them non-dimensional therefore requires recourse to their boundary and initial conditions. Those can usually be expected to define a representative length L and a representative velocity magnitude U. In d'Alembert's theorem (Fig. 13.1), for instance, the greatest diameter of the immersed body and the velocity magnitude of the uniform flow far upstream are possible choices for L and U, respectively. In the pipe flow problem of Section 19, natural reference quantities are the pipe diameter L and the ratio $4Q/(\rho\pi L^2) = U$ of the volume flow rate to the cross-sectional area. Different choices are also possible—for instance, the pipe radius may be selected in the place of the diameter—but such differences will be seen not to affect the considerations that follow. A transformation to nondimensional variables

(20.1) $$\mathbf{x}' = \mathbf{x}/L, \qquad t' = Ut/L, \qquad \mathbf{v}' = \mathbf{v}/U, \qquad p' = p/\rho U^2$$

then brings (4.5) and (19.3) into the nondimensional form

(20.2) $$\operatorname{div} \mathbf{v}' = 0$$

90

(20.3) $Dv'/Dt' = -\text{grad } p' + (Re)^{-1} \nabla^2 v'$,

where the derivatives are with respect to x' and t', and the nondimensional Reynolds number

(20.4) $Re = UL/\nu$

is the only physical constant. Any boundary and initial conditions expressing geometrical relations and conditions on v, p, and their derivatives are similarly made nondimensional by the transformation (20.1), so that they involve only parameters representing ratios of lengths or of velocity magnitudes. The precise meaning of "small viscosity" must therefore be

$$Re \to \infty.$$

In the most common physical circumstances the Reynolds numbers encountered lie between 10^4 and 10^8, and it is clearly desirable to describe such motions in terms of their limit as $Re \to \infty$.

In this limit (20.3) reduces to the nondimensional form of (12.3), and the limiting solutions of (20.2), (20.3) should therefore be solutions of the Euler system (4.5), (12.3). There is an important exception, however. If $|\nabla^2 v'|$ is very large compared with $|Dv'/Dt'|$, then (20.3) may differ radically from (12.3), even though Re^{-1} be very small. The proper conclusion is therefore that the viscous term distinguishing (20.3) from (12.3) is negligible in motions at large Reynolds number, *unless* $|\nabla^2 v'| \to \infty$ as $Re \to \infty$.

Conversely, we must generally anticipate that $|\nabla^2 v'| \to \infty$ somewhere in the fluid domain as $Re \to \infty$, for otherwise the term $Re^{-1} \nabla^2 v'$ would indeed become negligible in (20.3) so that the solution of (20.3) would tend to a solution of (12.3), which has been shown in Section 19 and Appendix 6 to be usually incompatible with Postulate IXb.

On the other hand, the sets in E^3 on which $|\nabla^2 v'| \to \infty$ as $Re \to \infty$ must be very restricted, in fact, must be of zero measure in some sense, for otherwise v' itself could not exist on the fluid domain in the limiting motion.

To look at it from a slightly more general point of view, let L_0 and L_1 denote linear differential operators with appropriate boundary conditions. The solution f of $L_0 f + \varepsilon L_1 f = 0$ for arbitrary small, positive ε is then called a *perturbation* of the solution f_0 of $L_0 f = 0$, because f is expected to be "nearly the same function" as f_0. If that turns out to hold in a straightforward sense, then the perturbation is called *regular*. But for instance, if L_1 is of higher order than L_0, the number of boundary conditions appropriate to L_1 is larger than that admissible for L_0, and f_0 cannot be expected to satisfy all the boundary conditions for L_1, even approximately. In particular, f and f_0 are likely to differ radically, however small ε may be, in a neighborhood of any boundary at which f satisfies more conditions than f_0. In this sense, slightly viscous fluid motion is a *"singular"* perturbation of ideal fluid motion, because the operator D/Dt' is of the first order, but ∇^2 is of the second order.

21. A Singular Perturbation Example

Before studying the system (20.2), (20.3) from this point of view, it is useful to discuss a simple example. Consider the problem [24] of finding the function $y(x) \in C^2[0, 1]$ satisfying

(21.1) $$\varepsilon \, d^2y/dx^2 + dy/dx = a$$

with positive constant $a < 1$ and boundary values

(21.2) $$y(0) = 0, \qquad y(1) = 1$$

for arbitrarily small positive ε. (We disregard that an exact solution can easily be written down, because our aim is to find an approach effective also in cases where the exact solution is unavailable or intransparent.)

Setting $\varepsilon = 0$ in (21.1) yields the "reduced equation"

$$dy/dx = a,$$

the general solution of which is

(21.3) $$y = f_0(x) = ax + b.$$

The constant b can be chosen so that f_0 satisfies either condition of (21.2), but not both. The most natural impulse, perhaps, is to satisfy the first of (21.2), so that $f_0 = ax$ (Fig. 21.1). To satisfy also the second of (21.2), y

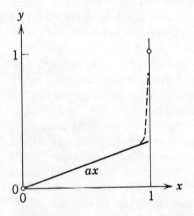

Fig. 21.1.

must then differ markedly from f_0 in some interval of x, and (21.1) shows this to be possible only if sufficiently large values of $y''(x)$ occur in that interval. In fact, $|y''(x)|$ must there grow beyond bounds as $\varepsilon \downarrow 0$.

The simplest conjecture is that an interval occurs near $x = 1$ (Fig. 21.1) in which $y(x)$ rises swiftly from $f_0(x)$ to its boundary value $y(1) = 1$. But such an interval must contain points at which $y''(x) > 0$ and $y'(x) > a$, and that is incompatible with (21.1).

The same difficulty precludes values of $y'(x)$ markedly exceeding a at any positive value of x independent of ε. Similarly, a maximum of $y(x)$ is precluded by the signs in (21.1). In short, the problem is simple enough to permit the strict conclusion, already at this stage, that $y(x)$ must, when ε is sufficiently small, increase monotonely from its boundary value $y(0) = 0$ to values differing arbitrarily little from those of (21.3) with $f_0(1) = 1$, i.e., from those of

(21.4)
$$f_0(x) = ax + 1 - a$$

(Fig. 21.2). This increase of $y(x)$, moreover, must occur in an interval of length shrinking to zero with ε, since y'' must fall below any bound in it, but y' must remain small compared with $|y''|$.

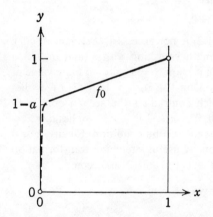

Fig. 21.2.

To understand the nature of the rapid transition of $y(x)$ from $y(0) = 0$ to $f_0(x)$, we need a magnifying glass that compensates for the shrinkage of the transition interval. In the present instance a transformation

(21.5)
$$\tilde{x} = \varepsilon^{-n}x, \qquad g(\tilde{x}) = y(x)$$

of the independent variable, with suitable real constant n, will turn out to be sufficient. It transforms the interval $[0, \varepsilon^n]$ of x into the interval $[0, 1]$ of $\tilde{x}$, whence the appropriate value of n would follow if it were known how fast the transition interval shrinks with ε.

This is not known to begin with, but it is possible to determine n from a slightly better understanding of the purpose of the *stretching transformation* (21.5). This will stretch the transition interval inadequately if $g(\tilde{x})$ still increases so fast that $|dg/d\tilde{x}|$ and $|d^2g/d\tilde{x}^2|$ (like $|dy/dx|$ and $|d^2y/dx^2|$) grow beyond bounds as $\varepsilon \to 0$. An essential condition on (21.5) is therefore that (if we indicate the implicit dependence on ε by a subscript) $g_\varepsilon(\tilde{x})$ and its first two derivatives with respect to $\tilde{x}$ tend respectively, as $\varepsilon \to 0$, to a function $g_0(\tilde{x}) \in C^2[0, \infty)$ and its first two derivatives. On the other hand, the stretching will be excessive if it shifts all appreciable variation of $g_\varepsilon(\tilde{x})$ to values of $\tilde{x}$ tending to infinity as $\varepsilon \to 0$; the magnifying glass then retains only an insignificant part of the transition in its field of view. In particular $dg_0/d\tilde{x}$ must vary markedly, because $g_0(\tilde{x})$ is intended to describe the transition of $y'(x)$ from its singularity at $x = 0$ to a value approximating a (Fig. 21.2). A second condition is therefore that $d^2g_0/d\tilde{x}^2 \not\equiv 0$.

For positive ε, however small, the transformation (21.5) brings (21.1) and (21.2) into the form

(21.6)
$$\frac{d^2g}{d\tilde{x}^2} + \varepsilon^{n-1}\frac{dg}{d\tilde{x}} = \varepsilon^{2n-1}a,$$

(21.7)
$$g(0) = 0.$$

The second boundary condition in (21.2) is not necessarily relevant to $g(\tilde{x})$, for if (21.5) maps an x-interval of length vanishing with ε (and including $x \doteq 0$) onto the unit interval of $\tilde{x}$, then it maps $x = 1$ onto no finite value of $\tilde{x}$. Now, in view of the first condition on (21.5), $n > 1$ in (21.6) implies $d^2g/d\tilde{x}^2 \to 0$ for all $\tilde{x}$, as $\varepsilon \to 0$, which contradicts the second condition. Similarly, in view of the first condition, $0 \le n < 1$ in (21.6) implies $dg/d\tilde{x} \to$ const for all $\tilde{x}$, which again contradicts the second condition. Finally, $n < 0$ is incompatible with the first condition. A useful stretching transformation can therefore occur only for $n = 1$, so that (21.5) and (21.6) become

(21.8)
$$\tilde{x} = x/\varepsilon, \qquad g(\tilde{x}) = y(x),$$

(21.9)
$$d^2g/d\tilde{x}^2 + dg/d\tilde{x} = \varepsilon a.$$

Setting $\varepsilon = 0$ in (21.9) gives a reduced equation quite different from that obtained earlier by setting $\varepsilon = 0$ in (21.1), and its solution satisfying (21.7) is

(21.10)
$$g_0(\tilde{x}) = c(1 - e^{-\tilde{x}})$$

with arbitrary constant c. This permits us to satisfy also the second condition of (21.2), which becomes $g(\infty) = 1$ in the limit $\varepsilon = 0$ (which we have already applied to (21.9)). Then $g_0(\tilde{x}) = 1 - \exp(-\tilde{x}) = 1 - \exp(-x/\varepsilon)$ (Fig. 21.3). This is a function satisfying both boundary conditions (21.2), but not the differential equation (21.1), even approximately, except for very small values of x. But if x_0 denotes the point of intersection of $f_0(x)$ and $g_0(\bar{x})$, we have found a continuous function,

$$f(x) = g_0(x/\varepsilon) \quad \text{for} \quad 0 \le x \le x_0$$
$$= f_0(x) \quad \text{for} \quad x_0 \le x \le 1,$$

satisfying both the boundary conditions (21.2) and satisfying (21.1) approximately, for sufficiently small ε, on each of the two subintervals. This function

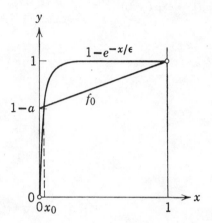

Fig. 21.3.

has a kink at x_0, however, and therefore fails to approximate the required solution $y(x) \in C^2[0, 1]$ to any satisfactory degree near x_0.

Prandtl first stressed that a satisfactory choice of the free constant c must "match" g_0 to f_0 so that there is an *open* x-interval in which both tend to the same smooth function as $\varepsilon \to 0$. This is necessary if f_0 and g_0 together are to approximate a smooth function in a sense including approximation of derivatives.

To see whether it is possible, let

(21.11) $\qquad \bar{x} = \varepsilon^{-\frac{1}{2}}x = \varepsilon^{\frac{1}{2}}\tilde{x}, \qquad h(\bar{x}) = y(x) = g(\tilde{x}).$

This transforms (21.1) into

$$\varepsilon^{\frac{1}{2}}\, d^2h/d\bar{x}^2 + dh/d\bar{x} = \varepsilon^{\frac{1}{2}}a,$$

and neither condition of (21.2) is necessarily relevant to $h(\bar{x})$, since this function is only intended to approximate $y(x)$ in an interval where both g_0 and f_0 are relevant, and therefore, in an interval not necessarily including either $x = 0$ or $x = 1$. Setting $\varepsilon = 0$ now gives

$$h(\bar{x}) = \text{const} = h_0,$$

and for sufficiently small ε this function is approximated arbitrarily closely by

$f_0(x)$ on an open interval such as $\frac{1}{2} < \bar{x} < 2$, if and only if $h_0 = 1 - a$, because $\bar{x} < 2$ implies $x \to 0$ as $\varepsilon \to 0$. It is similarly approximated by $g_0(\tilde{x})$ if and only if $c = h_0$, because $\bar{x} > \frac{1}{2}$ implies $\tilde{x} \to \infty$. (The same conclusion results from any choice $\bar{x} = \varepsilon^{-\alpha} x$ with $0 < \alpha < 1$.) Prandtl's idea therefore implies the *matching condition*

$$(21.12) \qquad\qquad c = 1 - a,$$

and then by (21.10),

$$(21.13) \qquad\qquad g_0(\tilde{x}) = (1 - a)(1 - \exp(-\tilde{x})).$$

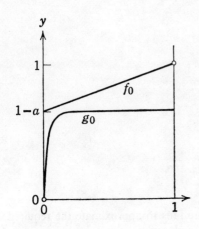

Fig. 21.4.

At first sight (Fig. 21.4) this looks even less satisfactory than the earlier choice (Fig. 21.3). But inspection of (21.13) and (21.4) shows that there is a smooth function

$$(21.14) \qquad y_0(x) = ax + (1 - a)(1 - e^{-x/\varepsilon}) \quad \text{for} \quad 0 \le x \le 1,$$

which, for sufficiently small ε, approximates (21.13) arbitrarily closely for all bounded $\tilde{x}$ and also approximates (21.4) arbitrarily closely for all x bounded away from zero. Thus $y_0(x)$ shows some promise of furnishing a *uniform* asymptotic approximation to the solution $y(x)$ of (21.1) and (21.2), that is, an approximation valid on the whole, closed interval $[0, 1]$ of x. Indeed, it is easy to verify that the exact solution is

$$y(x) = ax + (1 - a)(1 - e^{-1/\varepsilon})^{-1}(1 - e^{-x/\varepsilon}),$$

and $y_0(x)$ approximates this, together with all derivatives, with an error tending to zero with ε at all x in $[0, 1]$.

The limit, as ε reaches zero from positive values, of the solution of (21.1) and (21.2) is now seen not to be (21.3) for any choice of b. In fact, it is not a proper function of x. It equals (21.4) for $0 < x \le 1$, and at $x = 0$ it assumes all the values from zero to $(1 - a)$ inclusive.

In retrospect, the success of this heuristic approach to finding a uniform approximation to the solution of a singular perturbation problem, without obtaining the exact solution, is seen to be due mainly to the ideas of a stretching transformation and a matching condition. The former arises from the realization that a single independent variable x may not be adequate for a singular perturbation problem, since the limit of the solution need not be a proper function of a single variable. This indicates that simultaneous consideration of another independent variable, $\tilde{x} = \tilde{x}(x, \varepsilon)$, may be desirable. Two distinct limit processes then arise as $\varepsilon \to 0$. The first is the usual limit of the solution $y(x; \varepsilon)$ as $\varepsilon \to 0$ for fixed x. The second is the limit of $y(x; \varepsilon)$ as a function $y(x; \varepsilon) = g(\tilde{x}; \varepsilon)$, as $\varepsilon \to 0$ for fixed $\tilde{x}$. Then x varies with ε in some definite way as $\varepsilon \to 0$ in the second limit process; and similarly, $\tilde{x}$ varies with ε in the first limit process. When two distinct limits are thus considered simultaneously, they are sometimes called "inner" and "outer" limits.

The idea of the matching condition is expressed a little more generally and succinctly by the Prandtl Matching Principle, which eliminates the need to consider an additional, "intermediate" variable $\bar{x}$ in every instance. It may be put as follows. Consider a function of a variable x on $[0, 1]$ and of a parameter ε on $(0, 1)$ and define $\lim_A f$ as the usual limit of $f(x; \varepsilon)$ as $\varepsilon \downarrow 0$ for fixed x, and $\lim_B f$ as the limit of $f(x; \varepsilon) = g(\tilde{x}; \varepsilon)$ as $\varepsilon \downarrow 0$ for fixed $\tilde{x} = \tilde{x}(x, \varepsilon)$ [which also defines x implicitly as a function $x = x(\tilde{x}; \varepsilon)$]. Then if $\lim_A \tilde{x} = \infty$ and $\lim_B x = 0$, a necessary condition for the matching of asymptotic approximations $f_0(x)$ and $g_0(\tilde{x})$ to $f(x; 0)$ and $g(\tilde{x}; 0)$, respectively, is that $g_0(\infty) = f_0(0)$.

For our example (21.1, 2), the assumption of the preceding sentence applies to (21.8), and application of its conclusion to (21.10) and (21.4) implies immediately (21.12).

Naturally, these ideas are capable of refinement, and their great success [25] in mathematical physics has stimulated the growth of new branches of mathematics. The reader will find the relevance of stretching to viscous fluid motion confirmed by the solutions of Problems 19.1 to 19.4, all of which involve rapid transitions in boundary intervals of width shrinking to zero as the kinematic viscosity $\nu \to 0$.

Problem 21.1. Use stretching to find first approximations to all the roots of $\varepsilon x^3 + x^2 - 1$ for arbitrary small, positive ε.

Problem 21.2. Use stretching and matching to find a function $f(x; \varepsilon)$ likely to be a first approximation, uniformly on $[a, \infty)$, for arbitrarily small positive ε, to $y(x) \varepsilon C^1[a, \infty)$ such that

$$\varepsilon \, dy/dx + xy = 1, \qquad y(a) = b$$

with constants a and b such that $0 < a^{-1} < b$. Then confirm, by using the (stretched) differential equation to obtain an estimate for $(y - f)/\varepsilon$, that the error of the approximation is, everywhere on $[a, \infty)$, less than a fixed multiple of ε. (This is an instance in which the exact solution is not difficult to find but is so intransparent that the result is more quickly obtained by stretching, matching, and error estimation. The quite different problem arising on $(-\infty, a]$ is discussed in Appendix 21.)

Appendix 21

The preceding section, and those to follow, may prompt the reader to apply the ideas of stretching and matching to a variety of problems, and general experience indicates that this will prove fruitful in many instances. At the same time there have been occasions when investigators have been led astray by an excessive enthusiasm for this method. To balance the present account, it may therefore be desirable to add here a brief discussion of an illuminating example of Levey [26], which shows that a rapid transition in a boundary interval is not the only striking singular perturbation phenomenon, and that stretching-and-matching is not a panacea. Consider the function $y(x) \in C^1(-\infty, \infty)$ solving

$$(21.15) \qquad \varepsilon \, dy/dx + xy = 1$$

$$(21.16) \qquad y(a) = b$$

with constants a and b such that $0 < a^{-1} < b$, for arbitrarily small, positive ε. The reduced equation is $xy = 1$, and its unique solution, $y = f_0(x) = 1/x$, fails to satisfy (21.16). The intervals $(-\infty, a]$ and $[a, \infty)$ now present different problems. In the latter interval the problem responds readily to stretching and matching (Problem 21.2), but a new difficulty arises in the former interval. The reduced solution $f_0(x)$ is singular at $x = 0$. Equation 21.15, on the other hand, as a differential equation for a complex-valued function y of a complex variable x, has no singular point [27] in the complex plane, for any fixed $\varepsilon > 0$, however small. It follows [27] that all its solutions are entire functions and hence their real part y is a smooth function of x defined for all real x. In particular, $y(0)$ exists and, sufficiently close to $x = 0$, $y(x)$ and $f_0(x)$ must differ by an arbitrarily large amount!

A qualitative picture of $y(x)$ for very small, fixed $\varepsilon > 0$ is readily obtained from (21.15) and (21.16). Since $y(a) > 1/a$, the curve of $y(x)$ must pass through $x = a$ with large negative slope $y'(x)$. As x decreases from a, that curve must therefore rise very steeply (Fig. 21.5), but since $y(x)$ is defined for all x, it must still intersect the curve of $f_0(x) = 1/x$ at some $x = x_1 > 0$ (actually, with $y(x_1)$ exponentially large in $1/\varepsilon$, so that this part of the curve escapes from Fig. 21.5). Moreover, $y'(x_1) = 0$, since $f_0(x)$ is the zero-slope locus of all the solutions of (21.15). For $0 < x < x_1$, $y(x)$ must therefore have positive slope, and since $y(x)$ is defined for all x, its curve must cross the y-axis. For $x = 0$, its slope is $y'(0) = \varepsilon^{-1}$, by (21.15), and the curve of $y(x)$ must therefore continue to plunge steeply as x decreases from zero until it comes close to the other branch of $f_0(x)$, which it must necessarily intersect at some $x = x_2 < 0$, and again with $y'(x_2) = 0$.

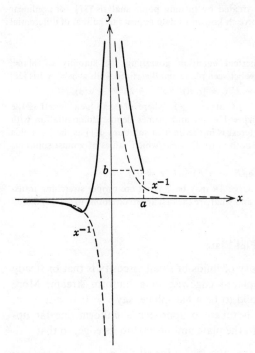

Fig. 21.5.

Now $f_0'(x) < 0$ for all x, so that $y(x)$ cannot cross its zero-slope locus from above to below with increasing x. For $x < x_2$, therefore, $y(x) < f_0(x)$ and $y'(x) < 0$. From (21.15), moreover, $y(x)$ must asymptote to $1/x$ as $x \to -\infty$. In fact, unless $|x_2|$ is large, $y''(x_2)$ is large, so that $y(x)$ must approach $f_0(x)$ very rapidly with increasing $x_2 - x$ (Fig. 21.5).

It will be apparent from the nature of $y(x)$ (Fig. 21.5) that the problem of describing this function quantitatively for $-\infty < x < a$, and especially for $-x_2 < x < a$, does not respond readily to stretching and matching as described by the Prandtl Matching Principle. When equations of the type (21.15) occur in applications, the most immediate aim tends to be the estimation of the "jump" of $y(x)$, i.e., of the difference between a and the zero of $y(x)$ (Fig. 21.5) or x_2 or, more rationally, the number c independent of ε such that $f_0(x)$ approximates $y(x)$ arbitrarily closely on $(-\infty, c)$ for sufficiently small ε. In Levey's example the exact solution

$$y(x) = e^{-x^2/(2\varepsilon)}\left[be^{a^2/(2\varepsilon)} + \frac{1}{\varepsilon}\int_a^x e^{u^2/(2\varepsilon)}\, du\right]$$

of (21.15) and (21.16) can be used to verify that $c = -a$.

Singular perturbations with singular interval (such as the neighborhood of $x = 0$ in Levey's example) occur in fluid dynamics in the theories of hydrodynamic stability [28] and of shock waves [29] (but not in the simple case of Section 37). The available evidence indicates that the explanation of the mechanism by which the wind generates water waves [30] also depends on such a singular layer. For linear ordinary differential equations such problems can be treated by turning point analysis [31]; for nonlinear differential equations the best approach known to date appears to be that of differential inequalities [32].

Problem 21.3. The Orr-Sommerfeld equation governing the stability of plane, viscous, incompressible, steady flow between plane, parallel, fixed walls at $y = \pm 1$ is [28]

$$(21.17) \qquad (U - c)(\psi'' - \alpha^2\psi) - U''\psi = (i\alpha R)^{-1}(\psi^{(iv)} - 2\alpha^2\psi'' + \alpha^4\psi)$$

with boundary conditions $\psi = \psi' = 0$ at $y = \pm 1$, where $\psi(y) \exp[i\alpha(x - ct)]$ is the perturbation stream function, $U(y) = 1 - y^2$, and primes denote differentiation with respect to y. The main physical interest is in the limit of solutions $\psi(y)$ as the Reynolds number $R \to \infty$ for fixed α and c, both in $(0, 1)$. Near which values of y must solutions of the reduced equation

$$(U - c)(F'' - \alpha^2 F) - U''F = 0$$

fail to approximate $\psi(y)$, however large αR may be ? What are proper stretching transformations near those values of y, and what are the first approximations to (21.17) there?

22.   Limit Equations for the Flat Plate

A basic problem of the theory of fluids of small viscosity is that of steady flow past a solid flat plate placed edgewise in a uniform stream. More precisely, the plate is understood to be a half-plane, say $y = 0, 0 < x < \infty$, $-\infty < z < \infty$, and the flow is taken to approach a uniform one far upstream, with velocity parallel to the plate and normal to its edge, so that

$$(22.1) \qquad \mathbf{v} \to U\mathbf{i} \quad \text{as} \quad x \to -\infty \quad \text{for all} \quad y, z,$$

where $U = \text{const} > 0$ and $\mathbf{i}$ denotes the unit vector in the direction of increasing x (Fig. 22.1). The symmetry of the problem indicates that there should be a two-dimensional flow, with velocity components $u, v, 0$ such that u and v are independent of z, and attention will be restricted to such a

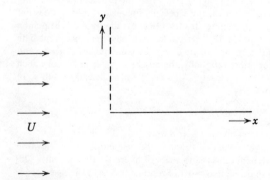

Fig. 22.1.

flow. The plate is understood to be impermeable, so that Postulate IXb imposes the boundary condition

(22.2) $u = v = 0$ at $y = 0$ for $0 < x < \infty$.

In view of the symmetry, moreover, there is no loss of generality in restricting attention to the half-space $y \geq 0$.

The boundary conditions (22.1) and (22.2) fail to define a length, so that it is not possible to apply the considerations of Section 20 immediately. The differential equations will therefore be taken in their dimensional form (4.5) and (19.3), i.e.,

(22.3)
$$\frac{\partial u}{\partial x} + \frac{\partial v}{\partial y} = 0,$$

(22.4)
$$u \frac{\partial u}{\partial x} + v \frac{\partial u}{\partial y} = \frac{-1}{\rho}\frac{\partial p}{\partial x} + \nu \left(\frac{\partial^2 u}{\partial x^2} + \frac{\partial^2 u}{\partial y^2}\right),$$

(22.5)
$$u \frac{\partial v}{\partial x} + v \frac{\partial v}{\partial y} = \frac{-1}{\rho}\frac{\partial p}{\partial y} + \nu \left(\frac{\partial^2 v}{\partial x^2} + \frac{\partial^2 v}{\partial y^2}\right)$$

(the last of (19.3) only confirms that the pressure p is also independent of z), and we shall consider their solutions in the limit as $\nu \downarrow 0$, waiting for an appropriate nondimensional formulation and an appropriate Reynolds number to emerge from the analysis.

For $\nu = 0$, (22.4) and (22.5) reduce to Euler's equation (12.3) for two-dimensional steady flow of ideal fluid, and (22.1) makes it likely that Lemma 13.1 will apply, so that the solution will be irrotational. But the problem is still indeterminate, for lack of further conditions at infinity. Clearly, however, an ideal flow of primary interest is the uniform flow

(22.6) $u \equiv U,$ $v \equiv 0$ for all $x, y,$

since experience with air and water suggests that a plate placed edgewise will not greatly disturb a uniform stream. Of course (22.6) fails to satisfy (22.2). The classical Flat-Plate Problem is defined as that of finding a solution of (22.1) to (22.5) differing from (22.6) only in a neighborhood of the plate the extent of which shrinks to zero as $\nu \downarrow 0$.

As noted in Section 20, if a solution exists, then it must possess a transition, or *boundary layer*, near the plate $y = 0$, in which $|\nabla^2 \mathbf{v}|$ grows beyond bound as $\nu \downarrow 0$. To study it, Prandtl introduced a stretching transformation

$$\zeta = y/\alpha(\nu)$$

of the normal distance from the plate. The problem may then be stated as that of finding a function $\alpha(\nu)$ tending to zero with ν and a solution $u(x, \zeta)$, $v(x, \zeta)$, $p(x, \zeta)$ of (22.1) to (22.5) (with $y = \alpha\zeta$) which has the limit (22.6), as

$v \downarrow 0$, for all x and all $y > 0$. If it exists, the Prandtl limit principle (Section 21) permits us to replace (22.6) by the matching condition

$$(22.7) \qquad \lim_{\zeta \to \infty} \lim_{v \to 0} u(x, \zeta) = U, \qquad \lim_{\zeta \to \infty} \lim_{v \to 0} v(x, \zeta) = 0,$$

$$\lim_{\zeta \to \infty} \lim_{v \to 0} p(x, \zeta) = \text{const} = p_0,$$

the last being implied by (22.6) and Postulate VII.

Observe that we have already rushed past a remarkable result. It has just been shown that, if a solution exists, then the limiting solutions as $v \downarrow 0$ must be, not the uniform flow (22.6), but the ideal flow which is this uniform flow together with a *vortex sheet* on the plate! The velocity must jump across this vortex sheet from the value $(U, 0)$ assumed at all interior points of the domain to the boundary value $(0, 0)$.†

Equation 22.3 implies the existence of a stream function $\psi(x, y)$ such that $u = \partial\psi/\partial y = \alpha^{-1}\,\partial\psi/\partial\zeta$, $v = -\partial\psi/\partial x$, and, say, $\psi(x, 0) = 0$. In the boundary layer u varies from 0 to U, by (22.2) and (22.7), and if $|u|$ is bounded independently of v, as is reasonable to expect, $\psi \to 0$ with α for all ζ. On the other hand, $\psi'(x, \zeta) = \alpha^{-1}\psi(x, y)$ is likely to remain a nondegenerate stream function as $\alpha \downarrow 0$. The proper stretching transformation should therefore be

$$(22.8) \qquad x' = x, \qquad \zeta = \alpha^{-1}y, \qquad \psi'(x', \zeta) = \alpha^{-1}\psi(x, y)$$

and

$$(22.9) \qquad \begin{aligned} u'(x', \zeta) &= \partial\psi'/\partial\zeta = \partial\psi/\partial y = u(x, y), \\ v'(x', \zeta) &= -\partial\psi'/\partial x' = -\alpha^{-1}\,\partial\psi/\partial x = \alpha^{-1}v(x, y), \end{aligned}$$

to satisfy (22.3). Since the purpose of a stretching transformation (Section 21) is to define functions that remain nonsingular, that is, retain all the derivatives required for the differential equations, even in the limit, a condition on $\alpha(v)$ is that ψ', u', and v' remain well defined as $v \downarrow 0$. It follows that $v(x, y) = \alpha v'(x', \zeta)$ will tend to zero with α for all ζ, and the second condition of (22.7) will be satisfied automatically by a proper choice of $\alpha(v)$.

The primes on x and u may now be dropped without danger of confusion. Substitution of the stretching transformation (22.8) into the remaining differential equations (22.4) and (22.5) then yields

$$(22.10) \qquad u\frac{\partial u}{\partial x} + v'\frac{\partial u}{\partial \zeta} = \frac{-1}{\rho}\frac{\partial p}{\partial x} + v\left[\frac{\partial^2 u}{\partial x^2} + \frac{1}{\alpha^2}\frac{\partial^2 u}{\partial \zeta^2}\right]$$

$$(22.11) \qquad \alpha^2\left[u\frac{\partial v'}{\partial x} + v'\frac{\partial v'}{\partial \zeta}\right] = -\frac{1}{\rho}\frac{\partial p}{\partial \zeta} + v\left[\alpha^2\frac{\partial^2 v'}{\partial x^2} + \frac{\partial^2 v'}{\partial \zeta^2}\right],$$

† The reader may note the sudden light shed on the relation between the concepts of free and bound vorticity (Section 8).

where v' is to be regarded as the function defined in terms of u by (22.9) rather than as an independent unknown.

Now let $v \to 0$. Then $\alpha(v) \to 0$ by hypothesis, and (22.11) shows p to tend to a function of x independent of ζ, whence, by (22.7), $p \equiv p_0$. If $v/\alpha^2 \to 0$, moreover, (22.10) and (22.11) tend to the Euler equations, and the stretching transformation fails to generate any new information. If $\alpha^2/v \to 0$, on the other hand, (22.10) implies $\partial^2 u/\partial\zeta^2 \to 0$ for all x and ζ, whence by (22.2), $u - a(x)\zeta \to 0$, and the matching condition (22.7) cannot be satisfied. A necessary condition for an appropriate stretching transformation is therefore that $\alpha^2/v \to \text{const} \neq 0$, and no generality is then lost in taking

$$\alpha^2 = v.$$

The equations (22.9), (22.10), (22.2), and (22.7) now reduce, as $v \downarrow 0$, to Prandtl's limit equations

(22.12) $\qquad u = \partial\psi'/\partial\zeta, \qquad v' = -\partial\psi'/\partial x,$

(22.13) $\qquad u\,\partial u/\partial x + v'\,\partial u/\partial\zeta = \partial^2 u/\partial\zeta^2,$

(22.14) $\qquad u(x,0) = v'(x,0) = 0 \qquad \text{for } 0 < x < \infty,$

(22.15) $\qquad \lim_{\zeta \to \infty} u(x,\zeta) = U \qquad \text{for } 0 < x < \infty.$

Although (22.3) to (22.5) are elliptic, these limit equations are parabolic; they are still nonlinear, but simpler, and a solution was found as follows.

Prandtl noticed that they still fail to define a reference scale for x or ζ, and this suggests a transformation from (x, ζ) to $(x'', \zeta'') = (ax, b\zeta)$ with constants a and b. Equations (22.12) and (22.13) then become

$$u = b\,\partial\psi'/\partial\zeta'', \qquad v' = -a\,\partial\psi'/\partial x'',$$
$$au\,\partial u/\partial x'' + bv'\,\partial u/\partial\zeta'' = b^2\,\partial^2 u/\partial\zeta''^2,$$

and (22.14) and (22.15) remain unchanged in form. The limit equations are therefore seen to be invariant under a transformation

$$\{x, \zeta, \psi'\} \to \{x'', \zeta'', \psi''\} = \{ax, b\zeta, b\psi'\}$$

if $b^2 = a$, for any $a \neq 0$. Accordingly, if $u(x, \zeta)$ is a solution, so is $u(b^2 x, b\zeta)$; the problem fails to distinguish between such sets of independent variables. But the solution cannot depend on the arbitrary constant b, and it must therefore depend on a combination of x and ζ that is independent of b. Any such combination must be a function of ζ^2/x, and such a solution must therefore be of the form $u(x, \zeta) = F(\zeta/x^{1/2})$. The quantity $\zeta/x^{1/2}$ still has dimension, but

(22.16) $\qquad \zeta(U/x)^{1/2} = y\left(\dfrac{U}{vx}\right)^{1/2} = \eta$

is nondimensional. The anticipated solution should thus have the form

(22.17) $\qquad u = Uf'(\eta), \qquad f' = df/d\eta,$

and since we have chosen $\psi(x, 0) = 0$, (22.12) and (22.9) imply

(22.18) $\qquad \psi = (\nu U x)^{1/2} f(\eta), \qquad v = \tfrac{1}{2}(\nu U / x)^{1/2}(\eta f' - f).$

By (22.13) to (22.15), finally, $f(\eta)$ must satisfy Blasius' equation

(22.19) $\qquad\qquad\qquad d^3 f / d\eta^3 + \tfrac{1}{2} f \, d^2 f / d\eta^2 = 0$

with boundary conditions

(22.20) $\qquad\qquad\qquad\qquad f(0) = f'(0) = 0$

and matching condition

(22.21) $\qquad\qquad\qquad\qquad \lim_{\eta \to \infty} f'(\eta) = 1.$

The solution of (22.19) to (22.21) was first computed by Blasius (1908). Existence and uniqueness of a solution in $C^\infty[0, \infty)$ was first proved by H. Weyl (1942); a simpler proof was given by Coppel [23], and a still simpler one, due to J. Serrin, is found in the appendix below. Figure 22.2 shows $f'(\eta) = u/U$ versus $\eta = y(U/\nu x)^{1/2}$.

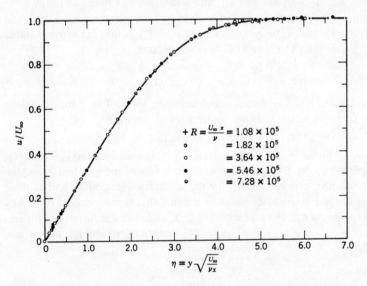

Fig. 22.2. Graph of $f'(\eta)$ for Blasius' problem, (22.19) to (22.21), with points representing experimental measurements of u/U at various Reynolds numbers $R = Ux/\nu$. From *Handbook of Fluid Dynamics*, V. L. Streeter (Ed.), Copyright 1961 by the McGraw-Hill Book Co. Used by permission of McGraw-Hill Book Co.

Appendix 22

Blasius' problem can be transformed into an initial value problem, for if $g(z)$ satisfies

(22.22) $g''' + gg'' = 0,$ $g(0) = g'(0) = 0,$ $g''(0) = 1$

and

(22.23) $\lim_{z \to \infty} g'(z) = 2k^2 > 0$

exists, then $f(\eta) = k^{-1}g(z)$, $\eta = 2kz$, satisfies (22.19) to (22.21). Now the fundamental theorem of ordinary differential equations [2] shows (22.22) to have a unique solution $g(z)$ on $[0, a)$ for some $a > 0$, and the existence of Blasius' solution can therefore be established by extending $g(z)$ to $[0, \infty)$ and confirming (22.23).

Since the differential equation may be written $(\log g'')' = -g$, (22.22) implies

(22.24) $g''(z) = \exp\left[-\int_0^z g(t)\, dt\right]$

on $[0, a)$, whence $g''(z) > 0$ on $[0, a)$ and again by (22.22), $g'(z) > 0$ and $g(z) > 0$ on $(0, a)$. But then (22.24) implies $g''(z) < 1$ on $(0, a)$, so that $0 < g(z) < z^2/2$ on $(0, a)$, whence again by (22.24), $g''(a)$, $g'(a)$, and $g(a)$ exist. The fundamental theorem therefore implies the existence of a unique solution $g(z)$ of (22.22) also on $[0, a + \delta)$ with $\delta > 0$. The same estimates, moreover, apply to it on $[0, a + \delta)$. Repetition of the same argument therefore establishes the existence of a unique solution $g(z) \in C^2[0, a]$ for every $a > 0$. The differential equation then implies $g(z) \in C^\infty[0, \infty)$.

It follows that the estimates obtained on $(0, a)$ hold for every $a > 0$. Accordingly $g(z)$ is monotone increasing on $[0, \infty)$, and in view of the initial values positive numbers b and c can be found such that $g(z) \geq b$ for $z \geq c$. From (22.24), therefore, $g''(z)$ *tends to zero exponentially* as $z \to \infty$.

$$\int_0^\infty g''(z)\, dz = g'(\infty)$$

thus exists, and it is positive because it has already been shown that $g''(z) > 0$ for $z \geq 0$.

To establish the uniqueness of Blasius' solution $f(\eta)$, suppose that another solution $f_1(\eta)$ also exists. Then

$$\log [f_1''(\eta)/f_1''(0)] = -\tfrac{1}{2} \int_0^\eta f_1(t)\, dt,$$

so that a solution $f_1 \not\equiv 0$ can exist only if a nonzero number $f_1''(0)$ exists. Denote $f_1''(0)$ by m^{-3}; then $g_1(z) = mf_1(2mz)$ also satisfies (22.22). But the solution of (22.22) is unique, and so $2m^2 = g'(\infty) = 2k^2$ and $f_1(\eta) = m^{-1}g(\eta/2m) = k^{-1}g(\eta/2k) = f(\eta)$.

Analytically $f(\eta)$ may be represented by its Taylor series

$$f = \alpha \frac{\eta^2}{2!} - \frac{\alpha^2}{2}\frac{\eta^5}{5!} + \frac{11\alpha^3}{4}\frac{\eta^8}{8!} - + \cdots$$

for small η, with $\alpha = f''(0)$. But this series has a finite radius of convergence. For large $\eta > 0$, $f(\eta)$ has an asymptotic expansion $f \sim \sum_0^\infty f_n(\eta)$ such that $|f_{n+1}/f_n| \to 0$ for all n, as $\eta \to \infty$. From (22.21) $f_0 = \eta - \beta$ with constant β, and from (22.19) $f_1'' + (\eta - \beta)f_1''/2 = 0$, so that

(22.25) $f \sim \eta - \beta + \gamma \int_\eta^\infty d\sigma \int_\sigma^\infty e^{-(\lambda - \beta)^2/4}\, d\lambda + \cdots$

with γ denoting another constant.

The constants α, β, and γ are best found by direct numerical integration of (22.22). Because of the exponential decay of $g''(z)$, the values of

$$\lim_{z \to \infty} g'(z) = \alpha^{-2/3}, \qquad \lim_{z \to \infty} [z - \alpha^{2/3}g(z)] = \alpha^{1/3}\beta$$

are readily determined, and it is found that

$$\alpha = f''(0) = 0.33206, \qquad \beta = \lim_{\eta \to \infty} (\eta - f(\eta)) = 1.7208.$$

Tables of $f'(\eta) = u/U$ are found in [21] and other monograph series.

23. Discussion of Blasius' Solution

We may now retrace our steps by substituting (22.16) to (22.18) into (22.4) and (22.5) to compare, for Blasius' velocity field, the magnitudes of the terms common to (22.4, 5) and (22.13) with the magnitudes of the terms of (22.4, 5) neglected by Prandtl; (22.3) is, of course, satisfied exactly. The substitution gives

$$\frac{2}{\rho U^2}\frac{\partial p}{\partial x} = \frac{1}{x}\left[2f''' + ff'' + \frac{v\eta}{2Ux}(\eta f''' + 3f'')\right],$$

$$\frac{2}{\rho U^2}\frac{\partial p}{\partial y} = \left(\frac{v}{x^3 U}\right)^{1/2}\left[f'' + \frac{\eta}{2}f'^2 - \tfrac{1}{2}ff' + \frac{v}{4Ux}(\eta^3 f''' + 6\eta^2 f'' + 3\eta f' - f)\right].$$

Since $f(\eta) \in C^\infty[0, \infty)$, the terms neglected by Prandtl are indeed seen to tend to zero, compared with any of the terms retained by him, as

(23.1) $\qquad v/(Ux) \to 0 \qquad$ for bounded η and x^{-1}.

For sufficiently small values of $v/(Ux)$ the solution of Blasius' problem (22.19) to (22.21) therefore satisfies (22.3) to (22.5) approximately for $0 < \eta < \eta_0$ and $x > x_0$ with arbitrary positive η_0 and x_0 independent of $v/(Ux)$. Hence Blasius' solution shows definite promise of furnishing an asymptotic approximation to a solution of (22.3) to (22.5) *in the limit* (23.1). (A proof of this long-standing conjecture has not yet been given.)

But for any given kinematic viscosity $v > 0$ (however small) and given $U > 0$, Blasius' velocity field is seen not to satisfy (22.4) or (22.5) even approximately, as $x \to 0$ for fixed η. Since that limit implies $y \to 0$ by (22.16), this failure of Blasius' solution is limited to a neighborhood of the *leading edge* $x = y = 0$ of the plate. The problem arising there has not yet been solved completely. It is connected with the particular formulation of Section 22 because the limit equation (22.13) does not have vectorial form and hence is not independent of the coordinate system chosen [33], but this does not affect the other features of Blasius' solution discussed below. It may also be relevant that the edge of a plate of zero thickness is physically not well defined. In any case, if y be fixed at any positive value, then $vx/U \to 0$ implies $\eta \to \infty$, so that $u(0, y) \to U$, by (22.21), and Blasius' solution blends into the uniform stream (22.6) without disturbing it upstream of the plate, and (22.1) is thus also satisfied. The failure of Prandtl's limit equations is therefore a local one, limited to a neighborhood of the leading edge which shrinks to the point $x = y = 0$ as $\nu \to 0$.

It emerges from all this that the relevant Reynolds number for the flat plate is the variable

$$(23.2) \qquad\qquad Ux/\nu = R_x.$$

"Fluid of small viscosity," in this case, must mean that R_x is large, and *every* Newtonian fluid therefore has "small viscosity" when x is large, or for fixed x, when the incident stream has a large velocity. To approach the inviscid limit of Newtonian fluid motion past the plate experimentally does not necessarily require a sequence of experiments with fluids of decreasing viscosity; it can be achieved equally well in a single experiment by moving the instrument further and further downstream from the leading edge, or by increasing the velocity of the incident stream.

The stretching transformation actually used in Section 22 is now seen to be

$$(23.3) \qquad\qquad \eta = (y/x)R_x^{\frac{1}{2}}, \qquad v' = (v/U)R_x^{\frac{1}{2}}.$$

The mathematical definition of "boundary layer" is seen to be the limit (23.1),

$$(23.4) \qquad\qquad R_x \to \infty \text{ with } \eta \text{ bounded},$$

and the definition of "ideal fluid flow" in this instance is seen to be the limit

$$(23.5) \qquad\qquad R_x \to \infty \text{ with } y/x \text{ bounded away from zero}.$$

In the latter limit the flow is uniform. The inviscid limit of Newtonian fluid flow past the plate comprises *both* limits; in the xy-framework it is the uniform flow together with a vortex sheet, and Blasius' solution predicts the approximate, internal structure of that vortex sheet.

Indeed, $\partial u/\partial y$ grows beyond bounds in the limit (23.4), but all the other first derivatives of u and v with respect to x and y remain bounded, according to (22.16) to (22.18). The vorticity components (7.2) are thus characterized by $|\omega_i/\omega_3| \to 0$ for $i = 1, 2$ and

$$R_x^{-\frac{1}{2}}\omega_3 \to -(U/x)f''(\eta).$$

In the limit (23.5), on the other hand, $|\boldsymbol{\omega}| \to 0$ everywhere.

It is also noteworthy that $\omega_3 \to 0$ exponentially with distance from the plate, in the boundary layer limit, because $f''(\eta) \to 0$ exponentially as $\eta \to \infty$ (Appendix 22). The asymptotic approximations in the two limits are thus seen to match exponentially smoothly.

Similarly, the shear stress components $\tau_{12} = \tau_{21} \sim \mu \, \partial u/\partial y = \tau$ in the limit (23.4), and for all other $i, j, |\tau_{ij}/\tau| \to 0$, by Postulate IXa. The non-dimensional shear stress

$$\tau/(\tfrac{1}{2}\rho U^2) = 2f''(\eta)R_x^{-\frac{1}{2}} \to 0,$$

and the motion therefore tends to an inviscid fluid motion, according to the naive definition of Section 12, also in the boundary layer limit.

It should be noted that the solutions of Problems 19.1 to 19.4 all possess a precisely analogous structure. These exact solutions of the equations of incompressible, Newtonian fluid motion therefore lend strong support to the conjecture that Blasius' solution furnishes an asymptotic approximation to the exact solution of the Flat Plate Problem.

The definition (23.4) shows the thickness of the boundary layer, measured normal to the plate, to be of the order of $xR_x^{-\frac{1}{2}}$, and because of the exponential approach of $f'(\eta)$ to its limit, $(U - u)/U$ is found to be practically negligible ($< 1\%$) already for $\eta > 5$, even though $u \to U$ only as $\eta \to \infty$. More definite measures of boundary layer thickness are provided by the percentage mass flow deficiency thickness, or briefly, *displacement thickness*,

$$\delta_1(x) \equiv \lim_{H^2 U/(vx) \to \infty} \int_0^H \left(1 - \frac{u(x, v)}{U}\right) dy,$$

and the *momentum* (flux deficiency) *thickness*,

$$\theta(x) \equiv \lim_{H^2 U/(vx) \to \infty} \int_0^H \frac{u(x, y)}{U} \left(1 - \frac{u(x, y)}{U}\right) dy.$$

Substitution of (22.16), (22.17), and (22.19) gives

$$\delta_1 = 1.7208 \, xR_x^{-\frac{1}{2}}, \qquad \theta = 0.6641 \, xR_x^{-\frac{1}{2}}$$

(see Appendix 22). All these thicknesses tend to zero as $v/U \to 0$ for fixed x, but not as $R_x \to \infty$ and $x \to \infty$; the vortex sheet diffuses (see Problem 19.1). The boundary layer is thus seen to be "thin" only in a qualified sense. For $R_x^{-1} = 0$ it lies between the plate and any straight line through the origin inclined to the plate at a nonzero angle. But for sufficiently large x it extends to arbitrarily large y. (The pipe flow problem of Section 19 is an instance in which consideration is restricted to values of x so large that the limit (23.5) has become empty.)

The tables of $f''(\eta)$ furnish the numerical values of the nondimensional shear stress $2\tau/(\rho U^2) = 2f''(\eta)R_x^{-\frac{1}{2}}$ and, in particular, the *skin friction coefficient*

$$c_f \equiv \tau|_{y=0}/(\tfrac{1}{2}\rho U^2) = 2f''(0) \, R_x^{-\frac{1}{2}}$$
$$= 0.6641 \, R_x^{-\frac{1}{2}},$$

except at the leading edge, where the Prandtl approximation is not valid. The aerodynamic drag per unit span exerted on any segment $0 < x < L$ of one side of the plate is $D_L = \int_0^L \tau_w \, dx$, where $\tau_w = \tau_{12}|_{y=0}$. On the assumption that τ_w is an integrable function of x, the drag may be computed from Blasius' solution in the limit $R_L = UL/\nu \to \infty$, since the interval of invalidity of the Prandtl approximation then tends to one of zero measure. This gives

$$R_L^{\frac{1}{2}} D_L \to 0.6641 \, \rho U^2 L,$$

so that the drag does tend to zero as $\nu \to 0$ for fixed U and L. D'Alembert's theorem 13.1 is thus seen to be not always misleading.

We are now in a position to clarify the remaining difficulty in the discussion of Postulate IXb in Section 18. By (18.4), the *nondimensional* slip velocity is

$$\frac{u(+0)}{U} = k_5 \frac{U}{c} c_f,$$

where k_5 is a numerical constant not differing greatly from unity, and c is the agitation velocity (Section 18). Now $U/c = k_6 M$, where [17] k_6 is another such constant, and M is the Mach number (Section 33), i.e., the ratio of the incident stream velocity U to the speed of sound of the fluid. For Blasius' solution the nondimensional slip velocity is therefore

(23.6) $\lim_{y \downarrow 0} (u/U) = k_7 M R_x^{-\frac{1}{2}}.$

Moreover, now that we consider definite problems with boundary conditions, we may state a rational, nondimensional *definition of inviscid motion of Newtonian fluid*. It is the limit

(23.7) $R_x \to \infty$ with M fixed

(with appropriate definitions of R_x and M). Postulate IXb follows for the Flat Plate Problem. By contrast,

$$R_x \to \infty \quad \text{with} \quad M R_x^{-\frac{1}{2}} \quad \text{fixed} \quad \neq 0$$

defines a *slip regime*, which represents a transition from continuum fluid dynamics to the *kinetic regime* defined as

$$R_x \to \infty \quad \text{with} \quad M/R_x \quad \text{fixed} \quad \neq 0.$$

In that limit, (18.2) shows l/x to be bounded, and bounded away from zero, so that the analysis of the motion requires the full tools of kinetic theory. *Free molecule flow*, finally, is defined as the limit

$$M/R_x \to \infty,$$

so that $x/l \to 0$ and molecular collisions are rare. We may now also give a definition of continuum fluid dynamics (as understood in this book): it is the limit $M/Re^{1/2} \to 0$, with definitions of Mach number and Reynolds number appropriate to the problem at hand.

To close the section, the remark should be added that the existence of a steady limiting solution as $R_x \to \infty$ does not imply its stability. In fact, the stability analysis [28] of Blasius' solution shows it to become unstable with respect to small, time-dependent disturbances at values of R_x not greatly exceeding 10^5. Experiment shows this instability to cause a transition to a flow, at values of R_x in excess of about 10^6, which is turbulent in the sense that its structure is so irregular that it can be analyzed only statistically. The limit (23.5) and a limit similar to (23.4) remain relevant nonetheless, and the limiting motion is one in which the time averages are uniform, together with a vortex sheet on the plate. But the structure of the vortex sheet and the mechanism of its diffusion are different, and e.g., the skin friction coefficient is $O((\log R_x)^{-1/2})$ [34] rather than $O(R_x^{-1/2})$.

Appendix 23

One of the best means of making boundary layers visible is offered by supersonic tunnels. The high velocity gradients in the boundary layer noted in Section 23 cause considerable viscous dissipation (Section 38) when the incident stream has a high velocity relative to the body, and the dissipation causes strong temperature gradients in the direction normal to the body surface. The discussion of Sections 22 and 24 shows the pressure gradient in that direction to be weak, and the same argument applies to compressible boundary layers. The equation of state (33.1) therefore shows that the strong temperature gradients are accompanied by similar density gradients, and those are made visible by a schlieren optic [35].

Figure 23.1 is a schlieren photograph of a steady, supersonic flow past a tetrahedron. It is mounted with a vertex pointing down and the left face parallel to the light rays of the optic. In short, the left edge of the black triangle is the image of this face of the tetrahedron, while the right edge of the black triangle is the image only of a tetrahedron edge.

The flow is upward, at Mach number 4, and the straight-line traces left and right of the black triangle are the schlieren images of shocks (Section 36). The schlieren edge is vertical, so that the optic shows horizontal density gradients, and only those. The orientation of the edge is such that a leftward increase of density casts a light image, and a rightward increase, a dark image. The light band bounding the left side of the black triangle and increasing slightly in thickness with distance in the direction of flow is the schlieren image of the boundary layer on the left face of the body.

24. Limit Equations with Pressure Gradient and Wall Curvature

The reader is justified in feeling skeptical about the wisdom of basing the sweeping conclusions of the last section on the one, special problem of flow past a flat plate. The present section is therefore devoted to showing briefly

Fig. 23.1. Schlieren photograph of a steady, supersonic flow past a tetrahedron. Crown Copyright Reserved.

that the Flat Plate Problem is not isolated and that the discussion of Section 23 has indeed general relevance if U and R_x are appropriately defined.

To begin, consider the flat plate $y = 0, 0 < x < \infty, -\infty < z < \infty$, placed in a nonuniform, steady, two-dimensional stream. For a more definite picture envisage the plate in its typical experimental setting in the test section of a wind tunnel. Such a tunnel is designed to produce a uniform flow in the empty test section, but by shaping the test section walls or simply attaching solid bodies to them, a great variety of nonuniform steady flows can be produced in an otherwise empty test section. Assume this to be done so that the shape of the test section remains symmetrical with respect to the plane $y = 0$; then the flow in the empty test section will be similarly symmetrical, and the plate will again be immersed edgewise in the stream.

As in Section 22, the search for the limiting solution of (22.3) to (22.5) as $\nu \downarrow 0$ begins with a consideration of the reduced solution, which is conventionally distinguished by a subscript e. For $\nu = 0$ (22.4) and (22.5) reduce to Euler's equation. It is again best to consider the flow before the plate is put in the tunnel. The tunnel has a settling chamber, upstream of the test section, in which the flow is normally found to be uniform even when the bodies are attached to the test section wall. By Lemma 13.1 the flow in the test section is therefore irrotational, and by Lemma 12.1

$$(24.1) \qquad p_e + \tfrac{1}{2}\rho(u_e^2 + v_e^2) \equiv \text{const.}$$

It will not be necessary here to know the details of this irrotational flow explicitly, but will suffice to consider $u_e(x, y)$ and $v_e(x, y)$ as known functions. Of course, they cannot be expected to satisfy Postulate IXb, but by symmetry

$$(24.2) \qquad v_e(x, 0) = 0 \qquad \text{for all} \quad x.$$

The Flat-Plate Problem may again be defined as that of finding a solution of (22.2) to (22.5),

$$(22.2) \qquad u(x, 0) = v(x, 0) = 0 \qquad \text{for} \quad 0 < x < \infty,$$

$$(22.3) \qquad \partial u/\partial x + \partial v/\partial y = 0,$$

$$(22.4) \qquad u\frac{\partial u}{\partial x} + v\frac{\partial u}{\partial y} = \frac{-1}{\rho}\frac{\partial p}{\partial x} + \nu\left(\frac{\partial^2 u}{\partial x^2} + \frac{\partial^2 u}{\partial y^2}\right),$$

$$(22.5) \qquad u\frac{\partial v}{\partial x} + v\frac{\partial v}{\partial y} = \frac{-1}{\rho}\frac{\partial p}{\partial y} + \nu\left(\frac{\partial^2 v}{\partial x^2} + \frac{\partial^2 v}{\partial y^2}\right),$$

that differs from $u_e(x, y)$, $v_e(x, y)$, $p_e(x, y)$ only in a neighborhood of the plate $y = 0, 0 < x < \infty$, such that the extent of this neighborhood shrinks to zero as $\nu \downarrow 0$.

This suggests the same stretching transformation, $\zeta = y/\alpha(\nu)$, as in Section

22, and the Prandtl limit principle (Section 21) now implies the matching conditions

(24.3) $$\lim_{\zeta \to \infty} \lim_{\nu \to 0} u(x, \zeta) = \lim_{y \to 0} u_e(x, y) = u_e(x, 0),$$

(24.4) $$\lim_{\zeta \to \infty} \lim_{\nu \to 0} v(x, \zeta) = \lim_{y \to 0} v_e(x, y) = 0,$$

(24.5) $$\lim_{\zeta \to \infty} \lim_{\nu \to 0} p(x, \zeta) = \lim_{y \to 0} p_e(x, y) = p_e(x, 0),$$

by (24.2), with $p_e(x, 0)$ and $u_e(x, 0)$ related by

(24.6) $$p_e + \tfrac{1}{2}\rho u_e^2 = \text{const}$$

on account of (24.1) and (24.2). It is customary to omit the explicit reference to $y = 0$ and to write simply $u_e(x, 0) = u_e(x), p_e(x, 0) = p_e(x)$ in (24.3) and (24.5). These matching conditions imply again that the limiting motion has a vortex sheet on the plate. Its strength is given by $u_e(x)$ and, in contrast to the situation of Section 22, this strength is not constant all over the plate, but it is again fully defined by the reduced solution.

It follows, just as in Section 22, that the stretching transformation must be (22.8), (22.9), that (24.4) is satisfied automatically, and that $\alpha^2 = \nu$ and $\nu^{-1}\, \partial p/\partial \zeta$ must remain bounded as $\nu \to 0$. Therefore $p(x, \zeta) \to p_e(x)$ by (24.5), and substitution of the stretching transformation (22.8), (22.9) into (22.2) to (22.5) and passage to the limit $\nu = 0$ give Prandtl's *Boundary Layer Equations*

(24.7) $$u = \partial\psi'/\partial\zeta, \qquad v' = -\partial\psi'/\partial x,$$

(24.8) $$u \frac{\partial u}{\partial x} + v' \frac{\partial u}{\partial \zeta} = \frac{-1}{\rho} \frac{dp_e}{dx} + \frac{\partial^2 u}{\partial \zeta^2},$$

(24.9) $$u(x, 0) = v'(x, 0) = 0 \qquad \text{for} \quad 0 < x < \infty,$$

(24.10) $$\lim_{\zeta \to \infty} u(x, \zeta) = u_e(x) \qquad \text{for} \quad 0 < x < \infty.$$

They differ from the flat-plate equations (22.12) to (22.15) mainly by the pressure gradient term in (24.8), which is, by (24.6) related to the velocity $u_e(x)$ of the reduced solution on the plate by

(24.11) $$-(1/\rho)\, dp_e/dx = u_e\, du_e/dx.$$

Conversely (22.12) to (22.15) are seen to be the special case of (24.7) to (24.11) corresponding to $u_e(x) = $ const. As in that special case (24.7) to (24.11) cannot be expected to furnish an approximation to the real fluid motion in a neighborhood of the sharp leading edge of the plate. But it is again reasonable to accept the long-standing conjecture that they promise an asymptotic approximation in the boundary layer limit (23.4), with $R_x = xu_e(x)/\nu$. Indeed, many of the considerations of Section 23 carry over to the solutions of (24.7) to (24.11) (Problem 24.1).

Next consider steady, two-dimensional flow past a realistic body of non-vanishing thickness. For definiteness we may envisage an airfoil spanning the test section of a wind tunnel with plane walls, so that the flow would be uniform in the absence of the airfoil. In any case the body is assumed to be solid, with impermeable surface. Its shape will define a fixed reference length, and hence a Reynolds number Re based on that, so that a nondimensional formulation (Section 20) is possible from the start. On the other hand, it is not possible to approach the problem in quite the same manner as before, for clearly there can be no Newtonian fluid motion which differs from the flow obtained in the absence of the body only by a perturbation which is small almost everywhere.

A rational approach is to *assume* that, for a fluid of small viscosity, the flow is one differing appreciably from that of an ideal fluid only in the vicinity of the body surface. More precisely, this amounts to the conjecture that the limiting solution, as $v \downarrow 0$, of (4.5), (19.3) and Postulate IXb is an ideal fluid flow together with a discontinuity surface (Section 9) coincident with the body surface.† We add the plausible assumption that this discontinuity is not one in the mass flow rate; then $\mathbf{v} \cdot \mathbf{n}$ must be continuous at the surface, and it must therefore (Section 9) be a vortex sheet, as in the flat plate flows discussed before and in the exact solutions of Problems 19.1 to 19.4. Since the body surface is impermeable, it now follows that the ideal flow must satisfy the condition $\mathbf{v} \cdot \mathbf{n} = 0$ on the vortex sheet.

Observe that this approach reflects the realization, which has emerged over the last fifty pages, that ideal fluid motion is one of a set of limit processes which *together* promise a useful description of real fluid motion. It is not a limit directly relevant to the motion very close to a solid boundary, and hence the physical boundary condition (Postulate IXb) is not directly relevant to this limit. To determine the ideal fluid limit uniquely requires instead a matching condition. Conversely, the condition (3.6) is now seen to play two quite distinct roles in fluid dynamics. On the one hand, it is a *part* of the physical boundary condition for real fluids. On the other hand, when (3.6) is applied to determine ideal fluid motion uniquely, then it is used as a matching condition on a vortex sheet, not as a physical boundary condition.

To return to the problem of flow past a body, the ideal flow limit will be irrotational in many cases. For instance, for the airfoil in the wind tunnel this can be deduced, by Lemma 13.1, from the observation that the airfoil has no measurable influence on the flow in the settling chamber upstream of the test section. Theorem 6.1 then indicates that the ideal flow may often be

---

† And perhaps extending downstream as a wake. But to divide the difficulties, it will be helpful to ignore this aspect in Sections 24 to 26.

uniquely determined by the matching condition (3.6) at the vortex sheet. This will be assumed in the following, so that the ideal flow may again be taken to be represented by given functions u_e, v_e, p_e satisfying (24.1).

To explore the internal structure of the transition represented in the limit by a vortex sheet, it is convenient to use curvilinear, orthogonal coordinates, which it is usual to denote again by x and y, such that the body surface is the line $y = 0$. A regular system of such coordinates is certain to exist, at any rate in a sufficiently small neighborhood of the body surface, provided that the body surface has no corner. A formulation of (4.5) and (19.3) in such coordinates is found in [34, p. 119]. It is convenient to measure x along the body surface from the stagnation point that must be anticipated near the rounded nose of a realistic body, and to begin with it is desirable to exclude a neighborhood of the point $x = y = 0$ from consideration. It is similarly desirable to exclude a neighborhood of the tail of the body.

Since y now measures normal distance from the body surface, Prandtl introduced the same stretching transformation $\zeta = Re^{1/2}y$, $v = Re^{-1/2}v'(x, \zeta)$ as before. On the assumption that (i) a twice differentiable solution $u(x, \zeta)$, $v'(x, \zeta)$, $p(x, \zeta)$ exists, in the limit $Re \to \infty$, for all ζ and for all x corresponding to points of the body surface, and (ii) the curvature $\kappa(x)$ of the body surface and its first derivative $d\kappa/dx$ are bounded (or at any rate, that $Re^{-1/2} \kappa \to 0$ and $Re^{-1} d\kappa/dx \to 0$) the component of (19.3) in the direction of increasing ζ is found [34] to reduce to

$$Re^{-1/2} \kappa u^2 = \partial p/\partial \zeta$$

as $Re \to \infty$. On the plausible assumption that u remains bounded, $\partial p/\partial y$ therefore also remains bounded in the boundary layer limit. But then, since the stretching implies a boundary layer thickness of order $Re^{-1/2}$, matching of the pressure in the two limits still requires $p(x, \zeta) \to p_e(x, 0)$ in the boundary layer limit. It is again usual to abbreviate $p_e(x, 0)$ and $u_e(x, 0)$ as $p_e(x)$ and $u_e(x)$, respectively.

Under the same assumptions the remaining equations of (4.5) and (19.3) are now found [34] to reduce precisely to the nondimensional form of (24.7) and (24.8) as $Re \to \infty$. The system (24.7) to (24.11) therefore may promise a first asymptotic approximation in the boundary layer limit (23.4), with $R_x = xu_e(x)/\nu$, also for realistic bodies with wall curvature, provided that they are sufficiently smooth. In fact, the wall curvature has no effect on this approximation, in contrast to the pressure gradient, which enters into the limit equations.

It may be noted in passing that the only effect of the nondimensional formulation (Section 20) upon (24.7) to (24.11) is the removal of the factor ρ^{-1}. It is more common to find the equations quoted in dimensional form, but with the stretching transformation reversed in the end, so that they become

$$\partial u/\partial x + \partial v/\partial y = 0,$$
$$u\,\partial u/\partial x + v\,\partial u/\partial y = -\rho^{-1}\,dp_e/dx + \nu\,\partial^2 u/\partial y^2,$$
$$u(x, 0) = v(x, 0) = 0 \quad \text{for all} \quad x,$$
$$\lim_{y/\alpha \to \infty} u(x, y) = u_e(x) \quad \text{for all} \quad x.$$

In the nondimensional form of this system, ρ^{-1} is removed and $\alpha = \nu^{1/2}$ is replaced by $\mathrm{Re}^{-1/2}$.

It is relevant, finally, to note that the solution of Problem 19.4 is an exact solution, not only of (4.5) and (19.3), but also of (24.7) to (24.11). It corresponds to $u_e(x) = \beta x$, $\beta = \text{const}$, and this is the first term of the power series of $u_e(x)$ with respect to the distance x along the body surface from any regular stagnation point of irrotational, two-dimensional, ideal fluid flow. With x measured from the front stagnation point of a smooth body, that solution therefore defines the proper limit of the solution of (24.7) to (24.11) as $x \downarrow 0$. This is important for the application of (24.7) to (24.11), since this system determines a unique solution only if $u(x_0, \zeta)$ is also specified for all ζ at some suitable value x_0 of x [36].

Problem 24.1. Show that for any solution of (24.7) to (24.11) with exponential approach to the limit (24.10), $xR_x^{-1/2}|\mathrm{curl}\,\mathbf{v}|$, $R_x^{1/2}c_f$, $R_x^{1/2}x^{-1}\,\delta_1(x)$, $R_x^{1/2}x^{-1}\theta(x)$, and $R_x^{1/2}M^{-1}\lim_{\zeta\to 0}(u(x, \zeta)/u_e(x))$ remain bounded in the boundary limit, where the skin friction coefficient, displacement thickness, and momentum thickness are defined respectively by

$$(24.12) \qquad c_f = (\tfrac{1}{2}\rho u_e^2)^{-1}\tau_{12}(x, 0),$$

$$(24.13) \qquad \delta_1(x) = \lim_{HR_x^{1/2}/x \to \infty} \int_0^H \left(1 - \frac{u(x, y)}{u_e(x)}\right) dy,$$

$$(24.14) \qquad \theta(x) = \lim_{HR_x^{1/2}/x \to \infty} \int_0^H \frac{u(x, y)}{u_e(x)}\left(1 - \frac{u(x, y)}{u_e(x)}\right) dy,$$

and the Mach number M is defined in Section 23 in terms of a representative value U of $u_e(x)$.

Appendix 24

Displacement effect. To obtain a more vivid physical interpretation of the displacement thickness (Section 23), note that solutions of Prandtl's limit equations (22.12) to (22.15) satisfy the first of the matching conditions (22.7) for all v, on account of (22.15), but the other matching conditions are satisfied only in the sense of (22.7), i.e., in the limit $R_x \to \infty$. For Blasius' solution, in particular,

$$\lim_{\eta \to \infty} v = \tfrac{1}{2}UR_x^{-1/2}\lim_{\eta \to \infty}(\eta f' - f) = \tfrac{1}{2}\beta UR_x^{-1/2} = 0.8604UR_x^{-1/2},$$

by (22.18), (22.21), and (23.2). Thus

$$\lim_{\eta \to \infty}(v/u) = d\delta_1/dx,$$

and for sufficiently small, positive R_x^{-1} and large enough η (but small y) Blasius' solution therefore satisfies to a close approximation the condition (3.6) for a solid body with impermeable surface given by the equation $y(x) = \delta_1(x)$. This suggests the conjecture that a refinement of the matching conditions (22.7) may be obtained by assuming the solution in the ideal flow limit (23.5) to be not a uniform flow but an ideal fluid flow past the body occupying the set $|y| < \delta_1(x)$ and satisfying (3.6) on its surface.

Since it is not quite a uniform flow, such a refinement would call for a corresponding correction to Blasius' solution by means of a solution of the limit equations of Section 24 for the appropriate functions $u_e(x)$, $p_e(x)$. Obviously these are first steps in a process of successive approximations in which progressively smaller corrections to the solutions in the limits (23.4) and (23.5) are computed in alternate steps. That such a scheme does furnish an asymptotic expansion with respect to $R_x^{-1/4}$ has been proved [37] for a solution of a linear system closely related to (4.5) and (19.3). But for the nonlinear system (4.5) and (19.3) no such proof is known at the time of writing.

25. Similarity Solutions

While the preceding sections have shed a great deal of light on the relation between ideal and real fluid motions, any impression that they explain it fully would be mistaken. They offer, in fact, less than half a loaf. The next major problem is difficult to present, and to help in broaching it some details of incompressible boundary layer theory will be discussed briefly in this and the next section. The sole purpose of these two sections is to extend the reasonably firm ground as far as possible toward the brink.

Blasius' solution is a member of the family of particular solutions of Prandtl's boundary layer equations (24.7) to (24.11) such that $u(x, \zeta)/u_e(x)$ depends only on a single variable $\eta(x, \zeta)$. A necessary condition for the existence of such "similarity" solutions can be shown [21] to be that $u_e(x)$ in (24.10) and (24.11) takes the form

$$u_e(x) = cx^m \quad \text{or} \quad u_e(x) = c \exp(\alpha x),$$

with constant c, m, α and with x measured from an arbitrary origin; of course, only non-negative values of x are admitted here for noninteger m, and for $m < 0$ only values with a positive lower bound. For $u_e(x) = cx^m$ (24.7) to (24.11) may be satisfied by

$$\frac{u(x, \zeta)}{u_e(x)} = \frac{df}{d\eta}, \qquad \psi(x, \zeta) = \nu R_x^{1/2} f(\eta),$$

(25.1)
$$\eta = (y/x) R_x^{1/2}, \qquad R_x = x u_e(x)/\nu$$

provided that

(25.2)
$$d^3f/d\eta^3 + \tfrac{1}{2}(m + 1)f\, d^2f/d\eta^2 = m(1 - (df/d\eta)^2),$$
$$f(0) = f'(0) = 0 \qquad \lim_{\eta \to \infty} f'(\eta) = 1.$$

For existence and uniqueness for this system, see [22, 23]. The case $m = 0$ is Blasius', and $m = 1$ and $\frac{1}{3}$ correspond, respectively, to Problem 19.4 and its axially symmetrical analog [38].

The solutions of (25.2) exhibit physical distinctions depending on the sign of the pressure gradient, which are reflected in the form of the *velocity profile*, meaning the curve of $u(x, \zeta)/u_e(x)$ versus η for fixed x (Fig. 25.1). From (24.8) and (24.9),

$$(25.3) \qquad \partial^2 u/\partial \zeta^2|_{\zeta=0} = \rho^{-1}\, dp_e dx$$

for any solution of Prandtl's equations, so that the velocity profile must have negative curvature at any point on the body surface where the pressure gradient is *favorable*, meaning $dp_e/dx < 0$ (so that the stream velocity u_e increases with x). Accordingly, the initial decrease of the vorticity magnitude, $\partial u/\partial y$, with distance from the wall, $\zeta = 0$, must be even more rapid there than for Blasius' solution. The similarity solutions for $m > 0$ exhibit this behavior; in fact, their velocity profiles lie altogether to the left of Blasius' (Fig. 25.1), the skin friction coefficient is larger, and the approach of u/u_e to unity at large η is even more rapid.

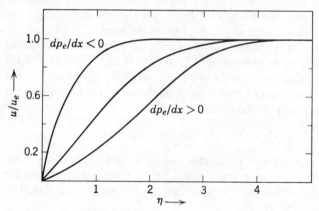

Fig. 25.1.

On the other hand, at any point where the pressure gradient is *adverse*, meaning $dp_e/dx > 0$, the velocity profile must have positive curvature, by (25.3). The largest vorticity magnitude then cannot occur at the wall, $\zeta = 0$, and since the vorticity must ultimately decay with increasing distance from the wall, the velocity profile must have a point of inflexion. The appropriate similarity solutions for $-0.0904 \le m < 0$ exhibit this behavior and lie altogether to the right of Blasius' (Fig. 25.1); the skin friction coefficient is smaller and the approach of u/u_e to unity at large η is slower. The skin friction coefficient decreases with m to zero at $m = -0.0904$. For still smaller m the velocity profile assumes negative values in a neighborhood of the wall, indicating a backflow there in which the fluid moves in the direction opposed to that of the general stream, and the physical relevance of the solutions is in doubt. For $m \geq -0.0904$ the velocity profile is monotone (Fig. 25.1).

The physical significance of the similarity solutions stems from the fact that a large class of solutions of Prandtl's equations (24.7) to (24.11) have the property of approaching an appropriate similarity solution asymptotically, with increasing distance x along the body surface. This has been partially explained [39] as follows. Assume that the pressure gradient is favorable and consider solutions for which the derivatives entering into (24.7) and (24.8) are continuous. It is reasonable to assume that the initial profile $u(x_0, \zeta)$ is positive for $\zeta > 0$, and it can then be shown [40] that $u(x, \zeta) > 0$ for $x \geq x_0$, $\zeta > 0$. By (24.7), the transformation to independent variables

$$\xi = x - x_0, \qquad \psi = \psi'(x, \zeta)$$

is therefore $1:1$ (even though singular at $\zeta = 0$). It was first introduced by Mises and transforms $x \geq x_0, \zeta \geq 0$ into $\xi \geq 0, \psi \geq 0$, and (24.8), (24.9), and (24.10) into

(25.4) $$\partial(u^2)/\partial\xi - \partial(u_e^2)/\partial\xi = u\,\partial^2(u^2)/\partial\psi^2,$$

(25.5) $$u(\xi, 0) = 0, \qquad \lim_{\psi \to \infty} u(\xi, \psi) = u_e(\xi), \qquad 0 \leq \xi < \infty,$$

by (24.11). Once $u(\xi, \psi)$ is known, ζ is obtained from

$$\zeta(\xi, \psi) = \int_0^\psi \frac{d\psi'}{u(\xi, \psi')}$$

by (24.7).

Now let $u_1(\xi, \psi)$ and $u_2(\xi, \psi)$ be two solutions of (25.4), (25.5) for the same pressure distribution $p_e(\xi)$ with initial values $u_1(0, \psi)$ and $u_2(0, \psi)$, respectively. Then $\phi(\xi, \psi) = u_2{}^2 - u_1{}^2$ satisfies

(25.6) $$\partial\phi/\partial\xi = u_2\,\partial^2\phi/\partial\psi^2 + \alpha\phi,$$

(25.7) $$\phi(\xi, 0) = \lim_{\psi \to \infty} \phi(\xi, \psi) = 0 \qquad \text{for} \quad \xi \geq 0,$$

with

$$\alpha = (u_2 + u_1)^{-1}\,\partial^2(u_1{}^2)/\partial\psi^2 = 2u_1{}^{-1}(u_2 + u_1)^{-1}\,\partial^2 u_1/\partial\zeta^2.$$

Assume that the Prandtl velocity profile, $u_1(x, \zeta)/u_e(x)$, of u_1 is everywhere concave downward, like the left-hand curve of Fig. 25.1, i.e., $\partial^2 u_1/\partial\zeta^2 < 0$ for all $x \geq x_0, \zeta > 0$ (this excludes even locally adverse pressure gradients, by (25.3)). Then $\alpha < 0$ for all $\xi \geq 0, \psi > 0$, and we can deduce that ϕ must assume its greatest and least values on the boundary of the domain $0 < \xi < \infty$,

$0 < \psi < \infty$ (parabolic maximum principle). Indeed, if ϕ did assume a value exceeding the boundary values at an interior point (ξ_0, ψ_0), then there would be a point (ξ_1, ψ_1), with $0 < \xi_1 \leq \xi_0$, at which the restriction of ϕ to the set $0 \leq \xi \leq \xi_0, 0 \leq \psi$, has a positive maximum, and at such a point $\phi > 0$, $\partial\phi/\partial\xi \geq 0$ and $\partial^2\phi/\partial\psi^2 \leq 0$, which is incompatible with (25.6) when $\alpha < 0$. A least value of ϕ at an interior point is excluded by an analogous sign contradiction.

Hence, if the initial data satisfy

$$\left|(u_2(0, \psi))^2 - (u_1(0, \psi))^2\right| \leq a^2$$

for some non-negative number a, then u_2 and u_1 satisfy the same inequality for all $\xi \geq 0, \psi \geq 0$, and since $au \geq 0$, also

$$|u_2(\xi, \psi) - u_1(\xi, \psi)| \leq a \qquad \text{for} \quad \xi \geq 0, \psi \geq 0.$$

But if the pressure gradient is strictly favorable, in the sense that $x\, dp_e/dx$ has a negative upper bound for $x \geq x_0$, then $u_e(\xi)$ grows beyond bound as $\xi \to \infty$. Hence if there is a solution u_1 with concave Prandtl profile, then the Mises profile, $u_2(\xi, \psi)/u_e(\xi)$, of *any* solution u_2 becomes asymptotically independent of the initial conditions. This is assured, in particular, for those favorable pressure distributions for which similarity solutions exist, since these are known to have a concave profile.

With refined estimates, this asymptotic result can also be extended [39] to the Prandtl profiles, and it can be shown that the Prandtl profile is concave, for sufficiently large $x - x_0$, if the pressure distribution is favorable and does not differ too greatly from one admitting a similarity solution.

In a more general manner, such asymptotic behavior is made plausible by the resemblance of Prandtl's equation to the classical diffusion or heat conduction equation,

$$\partial f/\partial x - \partial^2 f/\partial y^2 = g(x, y).$$

The analogy is displayed strikingly in Mises' form (25.4) of the equation, which is seen to be a nonlinear form,

$$\partial F/\partial\xi - u\, \partial^2 F/\partial\psi^2 = 0, \qquad F = u^2 - u_e^2,$$

of the homogeneous diffusion equation if u is everywhere non-negative. In general terms, the process described by the boundary layer equations may therefore be interpreted as a diffusion into the fluid of shear imposed by the boundary condition on the body surface. (Note also the analogy with Rayleigh's Problem 19.1.) Such a process may be expected, as in heat conduction, to smear out differences in initial conditions and to become gradually dominated by its internal mechanism and the amount of imposed shear, which is measured by $u_e(x)$. If $u_e(x)$ does not vary too rapidly, an approach toward a "self-preserving" form of the process, independent of initial conditions, is therefore plausible.

26. Momentum Integral

It is instructive to integrate Prandtl's limit equations (24.7) and (24.8) across the boundary layer. Begin by noting that (24.7) is equivalent to $\partial u/\partial x + \partial v'/\partial \zeta = 0$, whence

$$v'(x, h') = -\int_0^{h'} (\partial u/\partial x)\, d\zeta$$

and

$$v'\, \partial u/\partial \zeta = \partial(uv')/\partial \zeta + u\, \partial u/\partial x.$$

With h' taken independent of x, integration of (24.8) with respect to ζ from 0 to h' therefore yields

$$(26.1) \qquad \frac{\partial}{\partial x}\int_0^{h'} u^2\, d\zeta - u(x, h')\frac{\partial}{\partial x}\int_0^{h'} u\, d\zeta = \frac{-h'}{\rho}\frac{dp_e}{dx} + \frac{\partial u}{\partial \zeta}\Big|_0^{h'}.$$

If $u_H = u(x, h')$, $H = v^{1/2}h'$, $y = v^{1/2}\zeta$, and

$$\delta_1' = \int_0^h \left(1 - \frac{u(x, \zeta)}{u_h}\right) dy, \qquad \theta' = \int_0^h \frac{u}{u_h}\left(1 - \frac{u}{u_h}\right) dy,$$

(26.1) may, by (24.11), be written

$$(26.2) \qquad \begin{aligned} &v^{1/2}h'u_e\, du_e/dx + v(\partial u/\partial y)_{y=h} - \rho^{-1}\tau_w = \\ &\frac{d}{dx}[u_h^2(h - \delta_1' - \theta_1')] - u_h\frac{d}{dx}[u_h(h - \delta_1')], \end{aligned}$$

where $\tau_w = \mu(\partial u/\partial y)_{\zeta=0}$ is the shear stress on the body surface. Now let $v \to 0$ and $h \to 0$, but $h^2/v \to \infty$, in such a way that θ' and δ_1' tend to functions $\theta(x)$ and $\delta_1(x)$, that $v(\partial u/\partial y)_{y=h} \to 0$, and that $h(u_h - u_e(x)) \to 0$. This is certainly possible for any solutions of (24.7) to (24.11) sharing with the exact solutions of Problems 19.1 to 19.4, with Blasius' solution (Section 23), and with the relevant similarity solutions (Section 25), the property that the boundary layer limit (23.4) merges exponentially smoothly into the ideal flow limit (23.5). Then $u_h\delta_1' \to u_e(x)\delta_1$ and $\theta' \to \theta$ as $v \to 0$, where δ_1 and θ are the displacement and momentum thicknesses defined by (24.13) and (24.14), and (26.2) tends to Karman's momentum integral equation

$$(26.3) \qquad d(u_e^2\theta)/dx + u_e\, \delta_1\, du_e/dx = \rho^{-1}\tau_w,$$

or by (24.12),

$$d\theta/dx + (2\theta + \delta_1)u_e^{-1}\, du_e/dx = \tfrac{1}{2}c_f.$$

It shows that the boundary layer quantity of primary interest, the skin friction, depends only on $\theta(x)$, $\delta_1(x)$, and $u_e(x)$, but not on any local details of the velocity profile $u(x, \zeta)/u_e(x)$.

For most purposes, a boundary layer is therefore adequately characterized, at any given x, by the values of τ_w, θ, δ_1, and dp_e/dx. All their nondimensional combinations are products or ratios of

$$(26.4) \qquad l \equiv \frac{\theta\tau_w}{\mu u_e}, \qquad j \equiv \frac{\theta^2}{\mu u_e}\frac{dp_e}{dx}, \qquad H = \delta_1/\theta,$$

and any one solution of Prandtl's equations is therefore adequately characterized by three functions $l(x), j(x), H(x)$, and in fact, assuming x to be eliminable, by a pair of completely nondimensional functions $l(j)$, $H(j)$. The similarity solutions (Section 25) are an exception, since (25.1) and (24.12) to (24.14) show l, j, and H to depend only on m for these, so that each similarity solution is represented by a triplet of numbers l, j, H in this nondimensional formulation due to Thwaites. On the other hand, the family of similarity solutions, for $-0.0904 < m < \infty$, defines three functions $l(m)$, $j(m)$, $H(m)$, and upon elimination of m, two functions,

$$(26.5) \qquad l = l_s(j), \qquad H = H_s(j),$$

which, from the discussion of Section 25, are known to be asymptotically representative of a much larger class of solutions. This opens the possibility of integrating the momentum integral equation (26.3) in an asymptotic sense. Since $\tau_w = \nu l/(u_e\theta)$ and $u_e^{-1} du_e/dx = -\nu j/(u_e\theta^2)$, that equation takes the form

$$(26.6) \qquad (u_e/\nu) d\theta^2/dx = 2l + 2j(H + 2) \equiv L(j),$$

and since j is a known function of θ and x, when $u_e(x)$ is specified and (24.11) applies, substitution of (26.5) determines $\theta(x)$. The dependence of δ_1 and c_f on x is then obtained from (26.3) and (26.4).

Thwaites made a survey [41] of all approximate solutions of Prandtl's equations (24.7) to (24.11) known in 1948 and found that those appearing reliable were all approximated by (26.5) to quite a good degree, even for adverse pressure gradients, provided the skin friction was positive. This indicates that Serrin's results [39] should be extensible and that the similarity solutions should be asymptotically representative of an even much wider solution class than is presently established. On the other hand, since the significance of $l_s(j)$ and $H_s(j)$ is asymptotic, their substitution in (26.6) does not lead to a quantitative solution of the momentum integral equation for any boundary layer which is not actually a similarity boundary layer. What should be obtainable in that way is a qualitative guide to the development, with increasing x, of most normal boundary layers with positive skin friction. For such a purpose, use of the exact functions (26.5) is not mandatory, and

Thwaites pointed out that $L(j) \approx 0.45 + 6j$ is as good an approximation as any to all that was reliably known in 1948 (for a further improvement, see [21]). With that approximation, (26.6) integrates to

$$\theta^2 u_e{}^6 - \theta_1{}^2 u_1{}^6 = 0.45\nu \int_{x_1}^{x} [u_e(x')]^5 \, dx',$$

by (26.4), where θ_1 and u_1 denote the respective values of θ and u_e at the initial value x_1 of x. By (24.11), this may be written

(26.7)
$$\theta^2 u_e{}^6 = \theta_1{}^2 u_1{}^6 - 0.45\mu \int_{u_1}^{u_e} \frac{\lambda^6 \, d\lambda}{p_e'(\lambda)},$$

where $p_e'(u_e(x)) = dp_e/dx$.

The guide solution (26.7) exhibits strikingly the dominant influence of the pressure gradient, and especially of its sign, on the development of a boundary layer. If the pressure gradient is favorable then $u_e(x) > u_1$ for $x > x_1$, and because of the weight factor λ^6 the momentum thickness $\theta(x)$, and hence also $\delta_1(x)$ and $c_f(x)$, will usually depend only on the values of the pressure gradient in a neighborhood of x. This reflects, of course, the asymptotic result (Section 25) that the influence of the initial conditions is gradually lost, with increasing $x - x_1$, in a favorable pressure gradient. (In fact, if dp_e/dx tends to a negative value as $x - x_1 \to \infty$, then (26.7) predicts simply that $j = \theta^2 p_e'/(\mu u_e) \to -0.064$ or, roughly, $\theta^2 \, du_e/dx \to \nu/16$, as $u_e/u_1 \to \infty$.)

The situation is quite different for an adverse pressure gradient. Then $u_e(x) < u_1$ for $x > x_1$, so that $(u_1/u_e)^6$ increases rapidly with $x - x_1$ and the right-hand side of (26.7) usually depends only on the values of dp_e/dx in a neighborhood of the initial value x_1 of x. A boundary layer in an adverse pressure gradient therefore possesses a strikingly long memory for its initial conditions. Moreover, the momentum thickness increases rapidly as u_e/u_1 decreases.

In fact, $u_e(x)$ cannot usually be expected to remain positive in an adverse pressure gradient. If $x \, dp_e/dx \nrightarrow 0$ as $x - x_1 \to \infty$, then, by (24.11), $u_e(x)$ must vanish at a finite value of x, and by (26.7), θ cannot exist there, which contradicts the assumptions (Sections 22 and 24) on which Prandtl's boundary layer theory is founded.

Similarly, steady ideal fluid flow past a smooth, finite body will usually have a stagnation point at the rear of the body, and, by (26.3), the boundary layer approximation cannot then be expected to exist on the whole body surface.

27. Separation

It is as well to begin the discussion of such boundary layer breakdown in an adverse pressure gradient with the admission that our knowledge of it, at the time of writing, derives mainly from experimental observation. Three stages of the process can be distinguished in two-dimensional steady flow. The first is boundary layer development under an adverse pressure gradient. A second is characterized by negative skin friction. The last is complete "*breakaway*," in which the position of maximum shear is at a distance y_i from the body surface such that $\lim_{Re \to \infty} y_i > 0$, and the boundary layer is thus converted into a free shear layer.

The first stage (Sections 25 and 26) is a preparatory one, in which the position of maximum vorticity is at a positive distance y_i from the body surface, but $Re^{1/2} y_i$ is still bounded and the boundary layer limit still exists.

The second stage, where $c_f < 0$, involves negative values of u/u_e at sufficiently small y, (by (24.12) and (22.14) and because $\tau_{12} \sim \mu \, \partial u/\partial y$), and hence a "*backflow*" close to the body surface. It must begin at a "*separation point*" on the body surface at which $\partial u/\partial \zeta$ and the streamline pattern has a singular point (Fig. 27.1), in the sense that it is a point of confluence of three curves that are limits of streamlines. Two of them lie on the boundary surface, on either side of the separation point, and the third enters the fluid domain to separate regions of forward flow and backflow (Fig. 27.1).

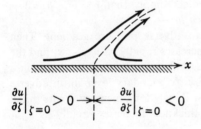

$$\left.\frac{\partial u}{\partial \zeta}\right|_{\zeta=0} > 0 \longrightarrow \longleftarrow \left.\frac{\partial u}{\partial \zeta}\right|_{\zeta=0} < 0$$

Fig. 27.1.

It is relevant to note that this singularity is not dependent on viscosity but is a near-universal kinematic feature of flows past solid bodies. The mere statement that the flow "passes" the body almost suffices to imply the existence of a "nose" and a "tail," respectively, characterized (in two-dimensional steady flow) by a point of attachment and a point of detachment at both of which the streamline pattern exhibits this same singularity. In three-dimensional flow many additional types of such kinematic singularities are possible, and an illuminating discussion of this aspect of separation is reported in [42]. At the same time Problem 19.4 and the next section make it clear that such kinematic attachment or separation need not imply a breakdown of the boundary layer limit concept. These examples, on the other hand, do not involve a clear-cut distinction between forward flow and backflow in the boundary layer, and in this respect they differ from the boundary layer separation first envisaged (Fig. 27.1).

But it is also uncertain that the occurrence of clear-cut backflow close to the body surface must always imply a breakdown of the boundary layer limit concept. For pressure distributions with a maximum, flows are not infrequently observed in which the backflow near the wall extends only over a short x-interval, between a point of kinematic separation and a point of kinematic reattachment, and in which any streamline that comes sufficiently close to the body surface upstream of that interval also comes arbitrarily close to it, downstream of the interval. In between, such a streamline passes over a "separation bubble," which may be very thin, and the boundary layer limit of Sections 23 and 24 may be applicable to some such bubbles.

That it cannot apply to all separation bubbles of bounded extent is made clear by the example of steady, two-dimensional flow "up a step" (Fig. 27.2). For appropriate conditions at infinity there exists a unique, irrotational,

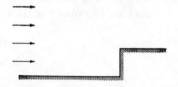

Fig. 27.2.

incompressible motion, continuous on the fluid domain, that satisfies (3.4) on the solid surface. It is easily computed in detail, by the help of the formulation of Section 6 and the Schwarz-Christoffel transformation [7]. It has a stagnation point at the concave corner, near which the flow approximates half that of Fig. 6.1. At the convex corner—near which the flow approximates that of the first of Fig. 6.7—it has a singularity of the velocity magnitude (and hence also of the pressure, by Bernoulli's Lemma 12.1). Both the adverse pressure gradient leading to stagnation and the unbounded, adverse pressure gradient at the singularity are inconsistent (Section 26) with the existence of a boundary layer limit (Section 24) complementing this potential flow to the inviscid limit of a Newtonian fluid motion.

It is intuitively plausible that a breakaway may occur well upstream of the concave corner and may be followed by reattachment at the convex corner, as sketched in Fig. 27.3, and experiment confirms that. The inviscid limit of the fluid motion is thus likely to be an ideal fluid motion together with a bound vortex sheet on the solid surface and a free vortex sheet of finite length (Fig. 27.3.). This flow nowhere resembles those of Figs. 6.1 and 6.7!

Fig. 27.3.

Examples abound of flows at large Reynolds number which must necessarily involve breakaway. For instance, the two-dimensional, steady potential flow (6.6) past a circular cylinder has a rear stagnation point (Fig. 6.2) and hence cannot represent even an approximation to a real fluid motion. The $\lim_{Re \to \infty}$ of real steady flow past a cylinder must therefore involve breakaway and formation of a pair of free vortex sheets. It is not easy to estimate the position of breakaway, nor is it intuitively obvious whether these free vortex sheets are of finite length, and, in fact, at the time of writing there is no general agreement in regard to the nature of this $\lim_{Re \to \infty}$.

The free vortex sheet arising from breakaway is quite analogous to the vortex sheet forming the boundary of a jet. Indeed, boundary layer separation resolves the ideal fluid paradox of breathing (Section 15). The streamlines of an ideal two-dimensional flow into, or out of, a duct are shown in Fig. 27.4.

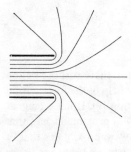

Fig. 27.4.

Outside the duct, far from its mouth, the potential approximates that of a source (6.7). For exhalation this ideal flow has a uniformly adverse pressure gradient on the outside of the duct wall, with a singularity of the pressure gradient at the lip, and there is therefore no boundary layer limit on the outside of the wall that can complement this ideal flow to the $\lim_{Re \to \infty}$ of a real fluid flow. The correct limit must be one in which the boundary layer inside the duct separates at the lip and a jet is formed. For inhalation, by contrast, the ideal flow of Fig. 27.4 has a uniformly favorable pressure gradient on the outside of the duct wall, and breakaway at the lip now leads to a jet *inside* the duct. Except for a small neighborhood of the lip, the streamline pattern outside the duct in Fig. 27.4 then represents the $\lim_{Re \to \infty}$ of a real fluid flow. We therefore reinhale only a small portion of the air just exhaled. This example shows strikingly how far real fluid motion may depart— even in the limit of zero viscosity—from the reversibility prediction (Section 15) of ideal fluid theory.

It will be plain enough, without further examples, that separation is a very common, and crucially important, feature of real fluid motions. The classical approach of Section 6, with its emphasis on continuous potential flow, though supported by satisfactory existence and uniqueness theorems (Appendix 6), is now seen to lead to results which can be very academic. But we must also retract the statement (Section 24) that (24.7) to (24.11) promise an asymptotic approximation in the boundary layer limit (23.4), and replace it with the weaker conjecture that they promise a transient asymptotic approximation in that limit. More precisely, with x denoting distance along the body surface from the nose, they promise an asymptotic approximation as $R_x = xu_e(x)/\nu \to \infty$ for fixed $(y/x)R^{1/2}$ and $x < x_*$, where x_* is some appropriate constant, but not an approximation uniform in x.

The problem of estimating x_* leads to a realization that steady fluid motions at high Reynolds numbers past solid bodies with kinematic attachment and separation may be divided roughly in two classes. The first consists of motions for which the ideal part of the inviscid limit is predictable independently of the details of the boundary layer limit, because the position of breakaway can be guessed a priori with convincing plausibility. Examples are airfoils at sufficiently small incidence for a Kutta-Joukowski condition (Section 6) to apply, and generally bodies with sharp edges, as in Figs. 4.1, 6.13, and 27.4, if a priori conditions are known which determine the ideal flow limit uniquely. The functions $u_e(x)$ and $p_e(x)$ in the boundary layer limit equations (24.7) to (24.10) are then also determined. In turn these functions, together with initial conditions furnished by Problem 19.4, determine the solution of (24.7) to (24.10) [36].

The second class consists of flows for which the ideal part of the inviscid limit depends strongly on details of the boundary layer limit, usually, because separation is caused by boundary layer development under an adverse pressure gradient (Section 26) on a smooth body surface. Examples are steady flow past a cylinder or a sphere, and flow up a step (Fig. 27.3). In the latter example the limit motion outside the free vortex sheet (Fig. 27.3) is again irrotational under appropriate conditions at infinity. This potential flow, and the pressure $p_e(x, y)$ in it, cannot be determined independently of the position of the free vortex sheet. That position, in turn, depends on the position of breakaway—which is determined by the development of the boundary layer under the adverse pressure gradient imposed by $p_e(x, y)$. For this flow, therefore, the coefficient functions $u_e(x)$ and $p_e(x)$ in (24.7) to

(24.10) cannot be determined independently of the solution of those equations, and this is typical of the second class of flows. The ideal flow and boundary layer parts of the inviscid limit of fluid motion are thus linked much more intimately than was tacitly assumed in Sections 24 to 26. Clearly, moreover, the flows of the second class are more common than those of the rather special, first class, to which Prandtl's approach applies directly. It should be added that in most practical instances of separation the flow is not two-dimensional or not steady.

28. Wake

It is helpful to round this chapter out with a brief discussion of the wake behind a solid body, in particular, the wake of a flat plate of finite length L placed edgewise in a uniform stream (Sections 22 and 23). It will be recalled that $u_e(x) = \text{const} = U$ in this case, and that the limiting, steady, two-dimensional flow of a Newtonian fluid, as $R_L = UL/\nu \to \infty$, can be conjectured confidently to be the uniform flow together with a vortex sheet on each side of the plate $y = 0$ for $L/N < x < L$, where N is any positive integer. That these vortex sheets cannot be expected to end at $x = L$ is already indicated by the conclusion of Section 1 that a line which is a boundary component of the fluid domain must extend from the plate.

This is clarified further by Blasius' solution (Sections 22 and 23), which shows that at large but finite R_x the vortex sheets are layers of strong shear in which arbitrarily small velocity magnitudes occur near $x = L$, $y = 0$. Since real fluid can be accelerated only gradually, by the smoothness convention, those shear layers must extend beyond the trailing edge of the plate, and since the lateral spread of the shear can similarly be only gradual, $(y/x)R_x^{\frac{1}{2}}$ must remain bounded in the shear layers over at least a small distance beyond $x = L$. The boundary layer limit (23.4) must therefore remain appropriate to a segment $L < x < L + \varepsilon$ of the wake formed by the shear layers. It follows that Karman's momentum integral equation (26.3) applies there.

But the whole flow is symmetrical with respect to $y = 0$, so that u is even and v odd in y, and u must therefore be continuous at $y = 0$. Thus $y = 0$, $x > L$, $|z| < \infty$ can be a discontinuity surface (Section 9) only with respect to velocity derivatives, and since it can only be a free discontinuity surface, it cannot support a shear force, whence $\partial u/\partial y$ must be continuous at it. The boundary condition (22.2) therefore gives way to

(28.1) $\qquad v = 0, \qquad \partial u/\partial y = 0 \qquad \text{for} \quad y = 0, x > L.$

Since $u_e(x) = \text{const}$, it follows from (26.3) that $d\theta/dx = 0$ for $L < x < L + \varepsilon$, so that $x^{-1}R_x^{\frac{1}{2}}\theta(x)$ actually decreases with increasing x, which indicates that the boundary layer limit (23.4) should remain valid for all positive

$ε$. (As in the earlier sections, we ignore the question of the stability of the steady solutions under discussion.) Then $θ(x) = $ const for all $x ≥ L$, and this invariant of the wake is directly related to the drag of the plate, which is

$$D = 2 \int_0^L τ_w(x) \, dx = ρU^2 θ(L),$$

by (26.3), where $θ = U^{-2} \int_{-∞}^∞ u(U - u) \, dy$ now stands for the full momentum thickness of the wake.

A simple exact solution for the wake in the limit (23.4) does not exist, but the qualitative nature of this limit becomes apparent upon transformation to the Mises variables (Section 25),

$$ξ = x - L, \qquad ψ = ψ'(x, ζ),$$

so that the governing equations take the form

(28.2) $$∂(u^2)/∂ξ = u \, ∂^2(u^2)/∂ψ^2,$$

(28.3) $$ζ(ξ, ψ) = \int_0^ψ \frac{dψ'}{u(ξ, ψ')},$$

by (25.4), for $ξ > 0$ and all $ψ$. From (28.1), moreover,

(28.4) $$u \, ∂u/∂ψ = 0 \qquad \text{for} \quad ψ = 0, ξ > 0.$$

Since $u > 0$ for $ψ > 0$ at $x = L$ (Fig. 23.1), it can be shown [40] that $u > 0$ for all $ψ$ at $x > L$, and (28.2) is a nonlinear variant of the diffusion equation. It follows that the velocity differences in the boundary layer at $x = L$ are evened out progressively by a diffusion process, as x increases, and we may therefore conjecture that $u(ξ, ψ) → U$ as $x/L → ∞$. This opens the way to a plausible asymptotic approximation of the final diffusion of shear, as $x/L → ∞$, based on the limit $(U - u(ξ, ψ))/U → 0$. Equation (28.2) then tends to the classical diffusion equation,

(28.5) $$∂(u^2)/∂ξ = U \, ∂^2(u^2)/∂ψ^2.$$

The known tendency toward asymptotic independence of initial conditions (Section 25) suggests a similarity solution of the form

$$u/U = f(η), \qquad η = ψ/(4Uξ)^{1/2},$$

but such a solution cannot represent the wake because it would imply a momentum thickness

$$θ(ξ) = U^{-2} \int_{-∞}^∞ u(U - u) \, dy = U^{-2} \int_{-∞}^∞ (U - u) \, dψ =$$
$$(4ξ/U)^{1/2} \int_{-∞}^∞ (1 - f) \, dη$$

increasing with x. Since $θ$ has been shown to be independent of x, any similarity solution must be such that $\xi^{\frac{1}{2}} \int_{-\infty}^{\infty} (U - u) \, d\eta$ is constant and must therefore have the form

(28.6) $\qquad U^2 - u^2 = U^2(v/\xi U)^{\frac{1}{2}} f(\eta), \qquad \eta = \psi/(4U\xi)^{\frac{1}{2}},$

since the asymptotic approximation is based on the limit $(U - u)/U \to 0$. Equations (28.5) and (28.4) then imply $f(\eta) = h \exp(-\eta^2)$, where $h = \text{const}$, and by (28.3) $\zeta = v^{-\frac{1}{2}} y \sim \psi/U$, so that our approximation is

$$U^2 - u^2 = hU^2 \left(\frac{v}{Ux}\right)^{\frac{1}{2}} e^{-y^2 U/(4vx)}$$
$$= 2U(U - u) + O((U - u)^2).$$

A further integration gives $h = U\theta/(\pi v)^{\frac{1}{2}}$. This approximation has the property that $|\partial^2 u/\partial x^2|/|\partial^2 u/\partial y^2| \to 0$ as $Ux/v \to \infty$, and we may therefore conclude, just as at the beginning of Section 23, that it promises to be an asymptotic approximation to a solution of the exact equations (4.5), (19.3) in the limit (23.4). It predicts a velocity deficiency in the wake of Gaussian shape in y, spreading like $(vx/U)^{\frac{1}{2}}$, with amplitude decaying like $R_x^{-\frac{1}{2}}$, and this agrees with experimental observation.

Analogous arguments apply to thin jets, and a similar spread and decay may be expected of many other free layers of strong shear in steady flows of fluids of small viscosity. For the basic example of time-dependent diffusion of a free vortex sheet, see Rayleigh's Problem 19.1.

In retrospect, the theory sketched in this chapter is seen to shed much light on the motion of real fluids, but also to raise profound new challenges for fluid dynamics. The reader will find it readily plausible that, even in the absence of solid bodies and separation, free discontinuity surfaces should be a common feature of the infinite Reynolds number limit of natural motions. Meterological fronts, for instance, are such surfaces; so are some ocean currents. The study of such questions is proceeding at a fast pace, but has not yet reached the completeness of boundary layer theory, and their discussion in this book will be limited to two more examples in the next chapter.

CHAPTER 5

Some Aspects of Rotating Fluids

29. Bjerknes' Theorem

Following common usage, the laws of fluid motion have been formulated in Chapters 2 and 3 for an absolute, inertial frame of reference (with occasional inclusion of gravitational effects) and their implications have then been compared with terrestrial experience. The next four sections will be used to clear up this inconsistency and to discuss briefly three classical results which give a slight taste of the remarkable effects of rotation on fluid motion.

To distinguish the experiences of a fixed and a rotating observer, let a subscript a denote quantities referred to an absolute, inertial frame of reference, and a subscript r, quantities referred to a frame rotating with angular velocity $\boldsymbol{\alpha}$ with respect to the absolute frame. Then if a vector $\mathbf{y}$ is fixed with respect to the rotating frame, its rate of change seen by the inertial observer is $d_a\mathbf{y}/dt = \boldsymbol{\alpha} \wedge \mathbf{y}$. Accordingly, if the vector is a function of time, $\mathbf{y}(t)$, in the rotating system, then its absolute rate of change is $d_a\mathbf{y}/dt = d_r\mathbf{y}/dt + \boldsymbol{\alpha} \wedge \mathbf{y}$. The absolute fluid velocity, defined by Postulate IV (Section 2), is therefore

$$\mathbf{v}_a \equiv \frac{\partial_a}{\partial t}\,\mathbf{x}(\mathbf{a},\,t) = \left(\frac{\partial_r}{\partial t} + \boldsymbol{\alpha} \wedge\right)\mathbf{x}(\mathbf{a},\,t),$$

and since the rotating observer sees only the change

$$\partial_r\mathbf{x}(\mathbf{a},\,t)/\partial t \equiv \mathbf{v}_r$$

in the position vector $\mathbf{x}$ of a point moving with the fluid, the respective fluid velocities for the two observers are related by

$$(29.1) \qquad \mathbf{v}_a = \mathbf{v}_r + \boldsymbol{\alpha} \wedge \mathbf{x}(\mathbf{a},\,t).$$

(We may forgo the subscript on $\mathbf{x}$, since we are at liberty to assume that the two frames coincide at the particular time t under consideration.)

It is natural for the rotating observer to define the circulation of a circuit C

according to the convention of Section 5, i.e., as $\Gamma_r(C) = \int_C \mathbf{v}_r \cdot d\mathbf{x}$, and the absolute circulation is then

$$\Gamma_a(C) \equiv \int_C \mathbf{v}_a \cdot d\mathbf{x} = \Gamma_r(C) + \int_C (\boldsymbol{\alpha} \wedge \mathbf{x}) \cdot d\mathbf{x}.$$

If a subscript $\perp$ denotes the component normal to the angular velocity $\boldsymbol{\alpha}$, then $(\boldsymbol{\alpha} \wedge \mathbf{x}) \cdot d\mathbf{x} = (\boldsymbol{\alpha} \wedge \mathbf{x}_\perp) \cdot d\mathbf{x}_\perp$, and therefore

$$\Gamma_a - \Gamma_r = \boldsymbol{\alpha} \cdot \left(\int_C \mathbf{x}_\perp \wedge d\mathbf{x}_\perp \right) = 2\alpha \Sigma(C_\perp),$$

where $\alpha = |\boldsymbol{\alpha}|$ and $\Sigma(C_\perp)$ is the area enclosed by the normal projection $C_\perp$ of C onto the equatorial plane, provided that the orientation of $C_\perp$ corresponding to that of C is related to $\boldsymbol{\alpha}$ by the right-hand screw rule. *In ideal fluid under a potential body force*, or in any other fluid motion to which Kelvin's theorem (Section 14) applies, $\Gamma_r(C) + 2\alpha\Sigma(C_\perp)$ *is therefore invariant for any circuit C moving with the fluid.*

This is the part of Bjerknes' theorem concerning the effect of rotation. As an illustration, consider a circuit C which is horizontal at latitude $\theta > 0$ (Fig. 29.1), has enclosed area A, and the greatest diameter of which is small com-

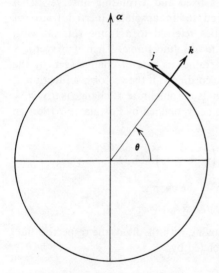

Fig. 29.1.

pared with the earth's radius. Then $\Sigma(C_\perp) = A \sin \theta$ approximately, and hence, a meridional translation, without any deformation, of the fluid mass containing C is sufficient to create relative circulation Γ_r.

30. Rossby Number

With a view to geophysical applications, attention is now restricted to steady rotation, so that $\boldsymbol{\alpha}$ is a fixed vector. To obtain the equations of motion of an incompressible Newtonian fluid in the rotating system, we begin by repeating the kinematical consideration of the preceding section. By (29.1), we obtain for the absolute fluid acceleration (Section 3) the expression

$$\frac{D_a \mathbf{v}_a}{Dt} \equiv \frac{\partial_a{}^2}{\partial t^2} \mathbf{x}(\mathbf{a}, t) = \left(\frac{\partial_r}{\partial t} + \boldsymbol{\alpha} \wedge\right)^2 \mathbf{x}(\mathbf{a}, t)$$

$$= \frac{\partial_r{}^2}{\partial t^2} \mathbf{x}(\mathbf{a}, t) + 2\boldsymbol{\alpha} \wedge \mathbf{v}_r + \boldsymbol{\alpha} \wedge (\boldsymbol{\alpha} \wedge \mathbf{x}),$$

and since $\boldsymbol{\alpha} \wedge \mathbf{x} = \boldsymbol{\alpha} \wedge \mathbf{x}_\perp$, where the subscript $\perp$ again denotes the equatorial component, $\boldsymbol{\alpha} \wedge (\boldsymbol{\alpha} \wedge \mathbf{x}) = -\alpha^2 \mathbf{x}_\perp$ with $\alpha = |\boldsymbol{\alpha}|$, and

(30.1) $$D_a \mathbf{v}_a / Dt = D_r \mathbf{v}_r / Dt + 2\boldsymbol{\alpha} \wedge \mathbf{v}_r - \alpha^2 \mathbf{x}_\perp.$$

The Momentum principle (Section 10) therefore implies for the rotating observer the equation of motion

(30.2) $$D_r \mathbf{v}_r / Dt + 2\boldsymbol{\alpha} \wedge \mathbf{v}_r = \mathbf{f} + \alpha^2 \mathbf{x}_\perp + \rho^{-1} \mathbf{P}$$

in place of (12.1), where $\mathbf{P}$ denotes the vector with components $\partial p_{ij}/\partial x_j$. It will be assumed here and in the next section that the body force $\mathbf{f}$ is the true gravitational force, and then

$$\mathbf{f} + \alpha^2 \mathbf{x}_\perp = \mathbf{g}$$

is just what we are used to calling the gravitational acceleration. The term distinguishing the relative equation of motion (30.2) from (12.1) is therefore the Coriolis force, $-2\rho\boldsymbol{\alpha} \wedge \mathbf{v}_r$. A simple implication is that an observer standing on the Northern Hemisphere, facing in the direction of $\mathbf{v}_r$ and looking at a body of fluid under no resultant force (i.e., with $\rho\mathbf{g} + \mathbf{P} = 0$), will not see it move straight ahead, but rather, will see it turn to the right.

The Coriolis force, moreover, is now seen to be the nonconservative force field responsible for the validity of Bjerknes', rather than Kelvin's, theorem in the relative motion of ideal fluid. The presence of such a force field makes irrotational motion, which plays so preponderant a role in classical, ideal fluid theory (and therefore also in our intuition), the exception rather than the rule. Moreover, not only are Lemma 13.1 and Corollaries 14.1 and 14.2 invalid for the relative motion, but also Helmholtz' Corollary 14.3 about convection of vortex tubes and lines. If the procedure of Section 14 is applied to (30.3) below, rather than (12.3), it leads to Helmholtz' equation in the form

$$D\boldsymbol{\omega}/Dt - (\boldsymbol{\omega} \cdot \mathrm{grad})\mathbf{v}_r = 2(\boldsymbol{\alpha} \cdot \mathrm{grad})\mathbf{v}_r + \nu \, \nabla^2 \boldsymbol{\omega}$$

for the relative vorticity $\boldsymbol{\omega} = \mathrm{curl} \, \mathbf{v}_r$, rather than (14.1). For the relative vorticity the classical convection of vortex lines is therefore accompanied not only by viscous diffusion of vorticity but also by its creation at the rate $2(\boldsymbol{\alpha} \cdot \operatorname{grad}) \mathbf{v}_r$.

To translate the constitutive equation (Section 17) into the rotating frame, let $\mathbf{y}(\mathbf{x})$ be any vector field. At any chosen time the absolute and rotating observers can use coincident coordinate systems, so that $\partial y_i / \partial x_j$, $i, j = 1, 2, 3$, have the same values for both. In particular, choose $\mathbf{y} = \boldsymbol{\alpha} \wedge \mathbf{x}$, so that $y_i = \varepsilon_{ijk} \alpha_j x_k$, where ε_{ijk} is the alternating tensor (Section 7); then

$$\operatorname{div} \mathbf{y} = \partial y_i / \partial x_i = \varepsilon_{ijk} \alpha_j \, \partial x_k / \partial x_i = \varepsilon_{iji} \alpha_j = 0,$$

$$\frac{\partial y_i}{\partial x_n} + \frac{\partial y_n}{\partial x_i} = \varepsilon_{ijk} \alpha_j \, \delta_{kn} + \varepsilon_{njk} \alpha_j \, \delta_{ki} = \alpha_j (\varepsilon_{ijn} + \varepsilon_{nji}) = 0,$$

so that the rotation does not affect div $\mathbf{v}$ or e_{ij}, and the constitutive equation (17.2) for an incompressible Newtonian fluid gives

$$\mathbf{P} = -\operatorname{grad} p + \mu \, \nabla^2 \mathbf{v}_r$$

for both observers. By (30.2), the equation of motion (19.1) therefore takes in the rotating frame the form

$$(30.3) \qquad \partial \mathbf{v} / \partial t + (\mathbf{v} \cdot \operatorname{grad}) \mathbf{v} + 2 \boldsymbol{\alpha} \wedge \mathbf{v} = \mathbf{g} - \rho^{-1} \operatorname{grad} p + \nu \, \nabla^2 \mathbf{v},$$

where $\mathbf{v}$ denotes the relative fluid velocity $\mathbf{v}_r$.

The boundary and initial conditions for (30.3) and (4.5) may be anticipated to define reference scales $U, L,$ and T for velocity, length, and time, respectively. (Usually, only two of these three scales will be independently defined, but this need not always be the case.) The equation of motion and these conditions together will then involve four nondimensional ratios of reference quantities: $L/(UT)$; the Froude number $U^2/(gL)$, where $g = |\mathbf{g}|$; the Rossby number

$$\begin{aligned} \operatorname{Ro} &= U/(\alpha L) \quad \text{if} \quad L/(UT) \quad \text{is bounded} \\ &= (\alpha T)^{-1} \quad \text{if} \quad UT/L \quad \text{is bounded}, \end{aligned}$$

which is a measure of the ratio of the relative fluid acceleration to the Coriolis acceleration; and the Ekman number

$$E = \nu/(\alpha L^2) = \operatorname{Ro}/\operatorname{Re}.$$

The other chapters of this book are thus seen to be concerned with the limit $\operatorname{Ro} \to \infty$ of terrestrial fluid motion, and, given α, that means motions of relatively short time scale T or L/U. For the earth $\alpha = 7.29 \times 10^{-5} \, \mathrm{sec}^{-1}$, so that the Rossby number is large for motions of a time scale not exceeding a few hours, and it is therefore not surprising that inertial fluid dynamics is an appropriate approximation for the discussion of many natural and technological fluid motions. On the other hand, the discussion of Chapter 4 will have alerted the reader to the fact that simple dimensional considerations of the type just indulged in are valid only for regular perturbation phenomena.

The other extreme, in which fluid motion is dominated by Coriolis effects rather than inertial effects, corresponds to the *geostrophic* limit Ro → 0. An example of this limit is provided by uniform relative motion, for which $\mathbf{v} \equiv \mathbf{U} = \text{const}$, so that $L^{-1} = T^{-1} = 0$ and (30.3) reduces to the geostrophic equation,

$$(30.4) \qquad 2\rho\mathbf{\alpha} \wedge \mathbf{U} + \text{grad}\, p = \rho\mathbf{g}.$$

A uniform, horizontal wind ("geostrophic wind") therefore blows *along* the isobars, rather than in the horizontal direction in which the pressure decreases most rapidly.

31. Ekman Layer

Many geophysical phenomena involve three disparate length scales, viz. the earth's radius, the much smaller depth of the fluid layer, and an intermediate scale characterizing the horizontal extent of the fluid motion. When the last scale is small compared with the earth's radius it is often convenient to use Cartesian coordinates based on a tangent plane (more precisely, a plane normal to $\mathbf{g}$). To cast (30.3) into such a form, let $\mathbf{i}, \mathbf{j}$, and $\mathbf{k}$ denote respectively the unit vectors pointing east, north, and vertically upward (Fig. 29.1). Then $\mathbf{\alpha} = \alpha\mathbf{k} \sin\theta + \alpha\mathbf{j} \cos\theta$ and the relative fluid velocity may be decomposed into horizontal and vertical components, $\mathbf{v}_h$ and w, respectively, by $\mathbf{v} = \mathbf{v}_h + w\mathbf{k}$, $\mathbf{v}_h \cdot k = 0$. Accordingly,

$$\mathbf{\alpha} \wedge \mathbf{v} = \alpha\mathbf{k} \wedge \mathbf{v}_h \sin\theta + \alpha w\mathbf{i} \cos\theta + \alpha\mathbf{j} \wedge \mathbf{v}_h \cos\theta,$$

and (30.3) may be split into a horizontal equation of motion,

$$(31.1) \quad \partial\mathbf{v}_h/\partial t + (\mathbf{v}_h \cdot \text{grad}_h + w\, \partial/\partial z)\mathbf{v}_h + 2\alpha(\mathbf{k} \wedge \mathbf{v}_h \sin\theta + w\mathbf{i} \cos\theta) = \\ -\rho^{-1}\,\text{grad}_h\, p + \nu\, \nabla^2\mathbf{v}_h,$$

and a vertical equation of motion,

$$(31.2) \quad [\partial w/\partial t + (\mathbf{v}_h \cdot \text{grad}_h + w\, \partial/\partial z)w]\mathbf{k} + 2\alpha\mathbf{j} \wedge \mathbf{v}_h \cos\theta = \\ \mathbf{g} - \rho^{-1}(\partial p/\partial z)\mathbf{k} + \nu\mathbf{k}\, \nabla^2 w,$$

where grad_h denotes the horizontal gradient operator, but ∇^2, the full Laplacian operator, and z is measured vertically upward.

The Coriolis force does not change the basic paradox of inviscid fluid dynamics; as Re → ∞, a boundary layer transition is needed to satisfy Postulate IXb at a solid surface. Ekman discovered an exact solution of the equations of relative motion, which satisfies Postulate IXb at an infinite, solid, tangential plane and approaches a uniform, horizontal motion with increasing distance from that plane. To derive this solution from (31.1) and

(31.2), let x and y denote Cartesian coordinates in the tangent plane, and $\mathbf{e}_1$, $\mathbf{e}_2$, u, and v, the corresponding unit vectors† and velocity components. The boundary conditions to be satisfied are

(31.3) $u = v = w = 0$ for $z = 0$, all x, y

and

(31.4) $u - U \to 0$, $v \to 0$, $w \to 0$ for all x, y as $z \to \infty$

with constant $U \neq 0$. The last condition is understood in the sense that all velocity derivatives also tend to zero, so that (31.1) implies

$$\text{grad}_h\, p + 2\alpha\rho U\mathbf{k} \wedge \mathbf{e}_1 \sin \theta \to 0$$

or

(31.5) $\partial p/\partial x \to 0$, $\partial p/\partial y \to -2\alpha\rho U \sin \theta$ for all x, y as $z \to \infty$.

Since all these boundary conditions are independent of t, x, and y, we may conjecture the existence of a solution such that $\mathbf{v} = \mathbf{v}(z)$, grad $p = \mathbf{f}(z)$. The incompressibility condition, div $\mathbf{v} = 0$, and (31.3) then imply $w \equiv 0$, and since grad$_h\, \partial p/\partial z = 0$, grad$_h\, p$ is independent of z and determined everywhere by (31.5). Moreover, $(\mathbf{v}_h \cdot \text{grad})\mathbf{v} \equiv 0$ and $\nabla^2 \mathbf{v} = \partial^2\mathbf{v}/\partial z^2$, so that (31.1) and (31.2) reduce to

(31.6) $2\alpha\mathbf{k} \wedge (\mathbf{v}_h - U\mathbf{e}_1) \sin \theta = \nu\, \partial^2\mathbf{v}_h/\partial z^2$,

(31.7) $2\alpha\rho\mathbf{j} \wedge \mathbf{v}_h \cos \theta + (\partial p/\partial z)\mathbf{k} = \rho\mathbf{g}$

for such a solution, of which the latter determines $\partial p/\partial z$, once $\mathbf{v}_h$ has been found from the former and (31.3) and (31.4). Note the contrast with the problem of Section 22; for $\alpha = 0$ (31.6) has no nontrivial solution such that $\mathbf{v}_h(0) = 0$ and $\mathbf{v}_h$ remains bounded as $z \to \infty$. But (31.6) may be written in components as

$$-2\alpha v \sin \theta = \nu\, \partial^2 u/\partial z^2,$$
$$2\alpha(u - U) \sin \theta = \nu\, \partial^2 v/\partial z^2,$$

and these may, in terms of $V = u - U + iv$, with $i^2 = -1$, and the variable Ekman number

$$\zeta^2 = z^2(\alpha/\nu) \sin \theta,$$

be recombined into

$$\partial^2 V/\partial \zeta^2 = 2iV$$

for $\alpha \neq 0$. Thus $V = A \exp(\lambda\zeta)$ with $\lambda^2 = 2i$ and $A = \text{const}$, and by (31.3) and (31.4), $A = -U$ and $\lambda = -1 - i$, whence

(31.8) $u/U = 1 - e^{-\zeta} \cos \zeta$, $v/U = e^{-\zeta} \sin \zeta$

(Fig. 31.1).

† They may, but need not, be identified with $\mathbf{i}$, $\mathbf{j}$, respectively.

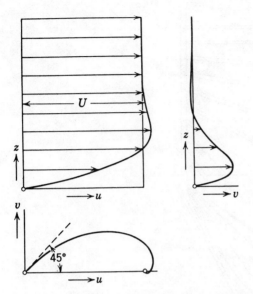

Fig. 31.1.   Profiles (31.8) of Ekman layer and Ekman spiral.

This solution describes a motion in a layer of thickness represented by $(\nu/(\alpha \sin \theta))^{1/2}$, which tends to zero with the viscosity. Note that it is an example of a three-dimensional boundary layer. The cross-flow component v normal to the uniform, geostrophic wind arises because the reduced, primary component u inside the boundary layer is not sufficient to balance the horizontal pressure gradient. The cross-flow leads to an angular velocity deviation β given by

$$\tan \beta = \frac{v}{u} = \frac{\sin \zeta}{e^{\zeta} - \cos \zeta},$$

so that $\beta \to \pi/4$ as $\zeta \downarrow 0$ and the deviation vanishes for, and only for, $\zeta = n\pi, \ n = 1, 2, \ldots$. The cross-flow is often illustrated by a plot of $v(\zeta)$ versus $u(\zeta)$, in which the solution is represented by Ekman's spiral (Figs. 31.1, 2).

It is relevant to note that (31.8) is an exact solution in the sense of Section 19, but from a geophysical point of view it is an approximate solution, because the earth's surface differs from an infinite tangent plane and on earth a uniform geostrophic wind cannot extend indefinitely in all directions. Of course, this scarcely diminishes the geophysical value of Ekman's solution in showing the effect of rotation on steady boundary layers. The solution even furnishes a good qualitative picture of the meteorological boundary

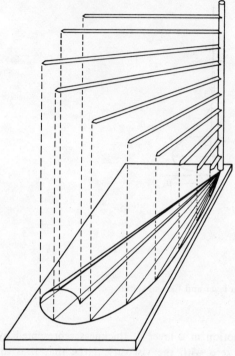

Fig. 31.2. Three-dimensional view of the velocity vector in Ekman's layer. From *General Meteorology* by H. R. Byers, Copyright 1937, 1944 by the McGraw-Hill Book Co. Used by permission of McGraw-Hill Book Co.

layer close to the ground, although that layer is thicker than (31.8) indicates and has a maximum angular deviation β of about 20°, rather than 45°, because the natural motion is turbulent.

It should also be noted that (31.1), (31.2) may often be approximated by simpler equations for the study of motions of horizontal scale large compared with the depth of the fluid layer. The vertical velocity may then become of secondary importance, so that (31.1) reduces approximately to

$$D\mathbf{v}_h/Dt + 2\alpha\mathbf{k} \wedge \mathbf{v}_h \sin\theta = -\rho^{-1}\operatorname{grad}_h p + \nu\nabla^2\mathbf{v}_h,$$

and this equation is used for many meteorological calculations. Frequently, moreover, $\alpha|\mathbf{v}_h|$ is small compared with $|\mathbf{g}|$, so that the pressure may be determined by the hydrostatic law $\partial p/\partial z = -\rho|\mathbf{g}|$. This simplified approximation to (31.1) draws attention to the fact that the relevant Rossby number for motions of horizontal scale small compared with the earth's radius is

$\text{Ro} = (T\alpha \sin \theta)^{-1}$ and thus depends on latitude. In particular, approximations based on the geostrophic limit $\text{Ro} \to 0$ must be anticipated to be nonuniform at the equator, $\theta = 0$.

32. Taylor-Proudman Theorem

A striking example of the geostrophic limit occurs when a solid body moves through a fluid in a rapidly rotating tank. We envisage an experiment in which the angular velocity of the earth is negligible compared with that of the tank and, in fact, disregard the earth's rotation. The equation of motion (30.2) relative to a frame rotating with the tank about a vertical axis at an angular velocity $\boldsymbol{\alpha} = \alpha\mathbf{k}$, where $\mathbf{k}$ is the upward, vertical unit vector and $\alpha > 0$, then becomes

$$D\mathbf{v}/Dt + 2\alpha\mathbf{k} \wedge \mathbf{v} = -g\mathbf{k} + \alpha^2\mathbf{x}_\perp + \rho^{-1}\mathbf{P},$$

where g denotes the usual gravitational acceleration, and $\mathbf{x}_\perp$, the component of the position vector normal to the tank axis. For an incompressible, Newtonian fluid $\mathbf{P} = -\operatorname{grad} p + \mu \nabla^2\mathbf{v}$, as in Section 30, so that the equation of motion is

$$(32.1) \quad D\mathbf{v}/Dt + 2\alpha\mathbf{k} \wedge \mathbf{v} = -\operatorname{grad}(\rho^{-1}p - \tfrac{1}{2}\alpha^2 r^2 + gz) + \nu \nabla^2\mathbf{v},$$

where r is the distance from the axis and z is the distance parallel to the axis, increasing upward.

Let L and U denote respectively length and (relative) velocity scales representative of the motion. Two reference scales, ρU^2 and $\rho\alpha^2 L^2$, are then available for the pressure; the safe procedure is to measure it in units of the largest available scale (since a limit process in which a scale ratio tends to zero might otherwise imply unbounded, nondimensional pressure and thus make the nondimensional equations meaningless). Since we wish to consider the limit $U/(\alpha L) \to 0$ in the following, p will therefore be measured in units of $\rho\alpha^2 L^2$. The nondimensional form of (32.1) is then

$$(32.2) \quad \text{Ro}\left(\frac{L}{UT}\frac{\partial\mathbf{v}}{\partial t} + (\mathbf{v}\cdot\operatorname{grad})\mathbf{v}\right) + 2\mathbf{k} \wedge \mathbf{v} =$$
$$-\operatorname{grad}\left[\frac{\alpha L}{U}\left(\frac{p}{\rho} - \tfrac{1}{2}r^2 + \frac{g}{\alpha^2 L}z\right)\right] + E \nabla^2\mathbf{v},$$

where $\text{Ro} = U/(\alpha L)$ and $E = \nu/(\alpha L^2)$, T is the time scale, and all the variables are now understood in units of their respective scales.

Taylor-Proudman theorem.   Under the circumstances just described let the Rossby number $\text{Ro} \to 0$, assuming that $E = \text{Ro}/\text{Re}$ and $\text{Ro}\, L/(UT)$ also tend to zero (which includes, in particular, the case of steady motion, in which $T^{-1} = 0$). Then the relative velocity $\mathbf{v}$ becomes independent of the vertical distance z on any open bounded point set S on which $\nabla^2 \mathbf{v}$ and $D\mathbf{v}/Dt$ tend to functions in $C^1(S)$.

For proof it suffices to note from (32.2) that the assumptions of the theorem imply

$$\frac{p}{\rho} - \tfrac{1}{2}r^2 + \frac{gz}{\alpha^2 L} \to \text{Ro } p_1(\mathbf{x}, t),$$

where $\text{grad } p_1 \in C^1(S)$. Thus $2\mathbf{k} \wedge \mathbf{v} + \text{grad } p_1 \to 0$ on S, by (32.2), and since a further differentiation is legitimate $\text{curl}(\mathbf{k} \wedge \mathbf{v}) \to 0$ on S. But $\text{curl}(\mathbf{a} \wedge \mathbf{b}) = (\mathbf{b} \cdot \text{grad})\mathbf{a} - (\mathbf{a} \cdot \text{grad})b + a \text{ div } \mathbf{b} - \mathbf{b} \text{ div } \mathbf{a}$, and since $\mathbf{k}$ is a fixed vector and the fluid is incompressible, (4.5) implies

(32.3) $$(\mathbf{k} \cdot \text{grad})\mathbf{v} \to 0$$

on S.

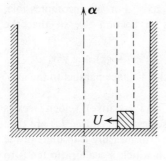

Fig. 32.1.

As a first corollary consider a tank with a plane bottom, perpendicular to the axis, across which a solid block of diameter L is pulled at a fixed, relative velocity U (Fig. 32.1). If the rotation rate of the whole assembly is increased so that $U/(\alpha L) = \text{Ro}$ becomes sufficiently small, and if the whole fluid domain is a set with the properties of S formulated above, the theorem predicts that the fluid column above the block moves together with the block, so that the fluid outside that column moves around the column, as if that were solid! This is because Postulate IXb is satisfied on the surface of the block and hence, by the theorem, also on the surface of the column. It follows, of course, that the whole fluid domain cannot be such a set S, but the prediction does not rule out that the interior and the exterior of the cylinder bounding the column are such sets, and the experiment confirms that.

As a second corollary consider a tank along the axis of which a sphere of radius L is pulled with constant velocity U (Fig. 32.2), and assume the tank

Fig. 32.2.

to be of unbounded extent in the axial direction. If the rotation rate is increased sufficiently, and if the whole fluid domain is a set with the properties of S, the theorem predicts that the fluid columns above and below the sphere must move with it, because the boundary condition on the equator of the sphere must, by the theorem, be satisfied on the whole vertical cylinder passing through the equator. Again this rules out that the whole fluid domain is such a set S, but not that the interior and the exterior of the cylinder are such sets. Indeed, the prediction was also confirmed by Taylor's experiment, at least in regard to the column ahead of the sphere, for Ro < 0.32.

Such predictions, of course, raise a host of fascinating questions about the geostrophic limit. The reader may, for instance, note the apparent non-uniqueness of the geostrophic limit of the equations governing incompressible, relative motion of Newtonian fluid. The equation of motion (32.2) reduces to

$$2\mathbf{k} \wedge \mathbf{v} + \operatorname{grad} p_1 = 0$$

on any set S (as defined in the theorem) and thus determines the variation of the pressure, once $\mathbf{v}(\mathbf{x}, t)$ is known. But the equations governing $\mathbf{v}$ are only (4.5) and (32.3), and thus imply only

$$\mathbf{i} \cdot \mathbf{v} = \partial \psi / \partial y, \qquad \mathbf{j} \cdot \mathbf{v} = -\partial \psi / \partial x, \qquad \mathbf{k} \cdot \mathbf{v} = w(x, y, t),$$

$$\psi = \psi_1(x, y, t) + zF(t)$$

for the Cartesian components of the relative velocity on a set S. Even in steady, relative flow, therefore, two functions remain to be determined by the boundary conditions. But the discussion of Chapter 4 shows the physical boundary conditions to be not directly relevant, because the geostrophic limit is of the singular perturbation type. Indeed, in contrast to the limit $\mathrm{Re} \to \infty$, which reduces the order of the differential equation (19.3) by one, the geostrophic limit reduces the order of (32.2) by two. Generally, therefore,

w, ψ_1, and F must be determined by matching conditions rather than boundary conditions, and Chapter 4 gives no adequate guide to these, because nonuniformities of the geostrophic limit must be anticipated from singularities, not only of $\nabla^2 \mathbf{v}$, but also of $D\mathbf{v}/Dt$. In analogy with the terminology of Chapter 4, these new nonuniformities are often called inertial layers or inertial jets [43].

In this connection, it may also be noted that the theorem shows it to be impossible to ignore the boundaries at the top and bottom of the tank by removing them to $z = \pm\infty$, as has been attempted in the discussion of the corollaries. Indeed, the flow must be anticipated to be very sensitive to the conditions at these boundaries. Consider, e.g., steady, relative flow in a tank with top and bottom bounding surfaces $z = h_i(x, y)$, $i = 1, 2$, respectively. The geostrophic limit (32.3) is then consistent with the matching condition (3.4) if and only if $h_2 - h_1$ depends only on $\psi(x, y)$ [43].

The brief remarks of this chapter show clearly that geophysical fluid dynamics poses a multitude of interesting questions in mathematical physics. Some of them are similarly relevant to magnetohydrodynamics, because of the analogy between the Coriolis force and the Lorentz force $\mathbf{H} \wedge \mathbf{v}$ on an ionized fluid under a magnetic field $\mathbf{H}$.

CHAPTER 6

Some Effects of Compressibility

33. Thermodynamic State

The restriction to incompressible fluid motion has been of help in the preceding chapters by excluding mathematical complications not germane to the issues, but it has necessarily led to a warped picture of fluid dynamics. An introduction to compressible fluid motion in the spirit of the earlier chapters would require a second book of similar length. It is worthwhile, however, to correct the balance partially by a sketch of two salient concepts of gas dynamics, namely, that of a shock and that of flow initiation by plane waves.

Postulates I to IX of Chapters 1 to 3 are clearly inadequate to this purpose, because they do not suffice to define the relation between the density and the other fluid-dynamical quantities. To remedy this requires the addition of the axioms of thermodynamics. But in view of the limited aim of this chapter, a more informal approach will be adopted, and thermodynamical notions will be introduced only to the minimum extent needed for the immediate discussion. Body forces and radiation will be ignored (except in Section 38), since they play no essential role in the processes considered.

Thermodynamics is concerned with substances many physical properties of which can be described in terms of a relatively small number of properties called *variables of state*. Examples of such variables are the *pressure* p, absolute *temperature* T, *density* ρ, and *entropy* S per unit mass. The primary concern of thermodynamics is therefore with relations between such variables, rather than with questions concerning their experimental definitions, and that point of view will also suffice for this chapter. It will be sufficient, moreover, to consider *pure substances*, defined as those for which the values of all variables of state are determined when the values of two among them are known. That implies, for any given pure substance, the existence of functional relations between any set of three variables of state, such as $S = S(p, \rho)$ or $\rho = \rho(p, T)$, which are called *equations of state*. An example is

143

the relation

(33.1) $$p = (c_p - c_v)\rho T$$

defining the class of pure substances called *perfect gases*; here $(c_p - c_v)$ is a thermodynamical constant, i.e., a quantity depending on the chemical nature of the gas but not on the variables of state. This example also illustrates how the value of the density can be found by means of an equation of state from the (more easily) measured values of pressure and temperature. Similarly, the entropy may in the following be regarded simply as a variable defined by an equation of state in terms of more familiar variables of state.

It will be helpful to select two particular variables of state for consistent use as independent variables in terms of which all the others are expressed, and those most convenient for our purposes are the density ρ and entropy S. Thus all equations of state will be envisaged in a form such as $p = p(\rho, S)$, $T = T(\rho, S)$. The notation $\partial p / \partial \rho$ then has an immediate meaning. This particular derivative is found to be positive for all physical real substances, and a further variable of state, $a(\rho, S)$, also positive, may therefore be defined by

(33.2) $$a^2 \equiv \partial p / \partial \rho.$$

It should be recalled from Postulate V (Section 4) that the density ρ can also take only positive values; the same holds of the absolute temperature T, by an axiom of thermodynamics; and the equations of state of all gases imply the same for the pressure p.

In many circumstances the behavior of real gases can be usefully approximated by that of the subclass of perfect gases for which another variable of state, the *specific heat at constant pressure*, $c_p(\rho, S)$, is constant. Since it differs for perfect gases by only a constant from the *specific heat at constant volume*, $c_v(\rho, S)$, these gases are called perfect gases with constant specific heat. They furnish particularly simple and useful illustrations of the general theory and will therefore be used as the main example throughout this chapter. In fact, it will not be worthwhile to give separate consideration to other perfect gases in this book, and for brevity the term "perfect gas" will in the following be taken to imply constancy of the specific heats. For such a gas the equation of state for the pressure can be shown (Appendix 35) to take the form

(33.3) $$p\rho^{-\gamma} \exp\left[-S/c_v\right] = \text{const},$$

where

(33.4) $$\gamma \equiv c_p/c_v > 1.$$

It follows from (33.2) and (33.1) that

(33.5) $$a^2 = \gamma p/\rho = (\gamma - 1)c_p T$$

for perfect gases.

All the notions just set out form part of the thermodynamics of equilibrium states, and it is not obvious that they can apply directly to the dynamical behavior of fluids. Experience shows, however, that a very large class of fluid motions involve only slight departures from *local* thermodynamical equilibrium, and this suggests the postulate that equilibrium thermodynamics can be *extended* to them. This may, in fact, be regarded as part of the definition of (classical) gas dynamics, to which the following account will be restricted. (All these matters are, of course, greatly illuminated by kinetic theory [18], but that is outside the scope of the present account.) The postulate has many facets, which will be introduced by and by at the points where they are needed in this chapter. In the first place, it is understood to imply that a local thermodynamic state is defined at each point in the gas at each time, and that these local states are related like thermal equilibrium states. The equations of state are therefore to be interpreted as functional relations between the functions of $p(\mathbf{x}, t)$, $T(\mathbf{x}, t)$, $S(\mathbf{x}, t)$, etc., representing the set of local states; these functional relations are independent of the motion and of the attendant variation of the state with $\mathbf{x}$ and t.

For a fluid at rest and in uniform state the thermodynamic definitions of density and pressure are identical with those given by Postulates V and VII. For a Newtonian fluid in motion the gas dynamical postulate will be interpreted here (consistently with kinetic theory [18]) in the sense that Postulates V and IXa supply those definitions. Inasmuch as a gas is a pure substance, its dynamical state may therefore be expected to be specified completely by the five scalar functions $\rho(\mathbf{x}, t)$, $S(\mathbf{x}, t)$, and $v_i(\mathbf{x}, t)$ $(i = 1, 2, 3)$. Postulates I to IX are taken to apply to them, and as in the earlier chapters the smoothness convention (Section 3) is understood unless the contrary is stated explicitly.

Just as it has been helpful in the preceding chapters to concentrate on incompressible fluid motion in order to gain an understanding of some salient effects of viscosity, it will be helpful now to confine attention to inviscid fluids as far as possible. In particular, this will be done in the rest of this section and in the next section. Postulate IXa is then reduced to the statement $p_{ij} = -p(\rho, S)\, \delta_{ij}$, and momentum conservation leads to Euler's equation (12.3), i.e.,

(33.6)
$$\rho D\mathbf{v}/Dt = -\operatorname{grad} p.$$

Postulate IXb must now be anticipated to be satisfied by the help of vortex sheets (Chapter 4), so that only the matching condition (3.4) need apply directly to the five functions ρ, S, and v_i. It should also be recalled from Appendix 18 that kinetic theory shows viscosity and heat conduction to be of similar importance in gases. Accordingly, an inviscid gas will automatically be understood to be devoid also of heat conduction. (The limit under consideration is, in fact, one in which $\mathrm{Re}^{-1} \to 0$ with fixed Mach number and Prandtl number $\sigma = \mu c_p/\lambda$.) The combination of equilibrium thermodynamics with the fluid dynamical postulates then implies

Theorem 33.1. In continuous motion of inviscid gas, entropy is convected with the fluid in its motion.

The reference to continuous motion is meant to express that the full force of the smoothness convention is not necessary; continuity of ρ, S, and $\mathbf{v}$ with respect to $\mathbf{x}$ and t is sufficient. The theorem is proved in Appendix 35 and has the immediate

Corollary 33.1. If a body of inviscid gas in continuous motion has uniform entropy at time $t = 0$, then its entropy remains uniform at all times.

Such gas motions are called *homentropic* and are of much practical interest, since many motions develop from a uniform thermodynamic state. The entropy then takes a value S_0 independent of $\mathbf{x}$ and t on the fluid domain, so that the equation of state gives $p = p(\rho, S_0)$ as, effectively, a function only of ρ. Now the motion of inviscid gas is governed by (33.6) and (4.3), i.e.,

$$(33.7) \qquad D\rho/Dt + \rho \operatorname{div} \mathbf{v} = 0,$$

expressing conservation of mass, and these are a system of four scalar equations for the five unknowns, ρ, p, and v_i ($i = 1, 2, 3$). For homentropic motion, the equation of state therefore provides the fifth equation needed to complement the system.

For steady flow of this type, it is now easy to see in what sense incompressible flow is a limit of compressible flow. From (33.2) and (33.6), since the flow is homentropic,

$$\operatorname{grad} \rho = a^{-2} \operatorname{grad} p = -a^{-2}\rho(\mathbf{v} \cdot \operatorname{grad})\mathbf{v},$$

and therefore

$$|\rho^{-1}\mathbf{v} \cdot \operatorname{grad} \rho| < M^2 |\operatorname{grad}|\mathbf{v}| \,|,$$

where $M = |\mathbf{v}|/a$ is called the (local) *Mach number*. The discussion of Chapter 4 has shown that boundedness of $|\operatorname{grad}|\mathbf{v}|\,|$ on the fluid domain is a necessary condition for a fluid motion to be regarded as inviscid. Hence, if $M \to 0$ uniformly on the fluid domain, ρ becomes constant along streamlines and (33.7) tends to the incompressible condition (4.5). It is significant that the bound for $|\operatorname{div} \mathbf{v}|$ is proportional to M^2, and in fact, appreciable effects of compressibility are rarely encountered in steady flows in which M nowhere exceeds $\frac{1}{2}$. (The main exception is free convection, in which buoyancy is the very cause of the motion.)

Appendix 33

Steady homentropic flow in ducts. The simplest application of the notions just developed is to steady, one-dimensional motion, and interesting results are obtained when this is extended to ducts in which the flow is not strictly one-dimensional, as in Sections 4 and 11. As noted there, this involves application of the principles governing local properties of the flow to their averages over the cross-section of the duct, and also neglect of the differences between products of such averages and the averages of the products. The justification of such approximations requires more detailed analysis of the actual, three-dimensional flows, and this problem has received relatively little attention. More indirect evidence, however, indicates that such a quasi-one-dimensional model can furnish a useful approximation to a surprisingly large class of flows [44]. This applies particularly to ducts which are straight and possess a cross-section changing only slowly in shape and area with distance x along the duct axis.

If it be also assumed that the fluid is inviscid and that the flow is continuous and comes from a reservoir in which the motion and thermodynamic state are uniform, then it follows from Corollary 33.1 that the flow is homentropic. The equation of motion (33.6) therefore reduces to

(33.8) $\qquad 0 = u \, du/dx + \rho^{-1} \, dp/dx = u \, du/dx + \rho^{-1}a^2 \, d\rho/dx,$

by (33.2), where $u(x)$ denotes the appropriate average of the velocity magnitude over the cross-section. As at the end of Section 33, the neglect of viscosity will for consistency be taken to imply also that $|du/dx|$ is bounded. Conservation of mass is expressed by (4.6), i.e.,

$$\rho u A(x) = \text{const},$$

where $A(x)$ denotes the area of the cross-section. Elimination of ρ gives

(33.9) $\qquad \dfrac{1}{A}\dfrac{dA}{dx} = (M^2 - 1)\dfrac{1}{u}\dfrac{du}{dx},$

where $M(x) = u/a$ is again the local Mach number. If $M < 1$, the flow is called locally subsonic (on account of the interpretation of the variable of state a derived in the next section) and u is seen to increase with x if and only if A decreases with x. As noted in Section 4, acceleration of the fluid in steady flow therefore requires a duct which narrows in the direction of the stream. If $M > 1$, by contrast, the flow is called locally supersonic and u is seen to increase with x if and only if A also increases with x. Acceleration of the fluid in steady supersonic flow therefore requires a duct which widens in the stream direction!

This result is easily extended for a perfect gas to

Lemma 33.1. If $M(x_0) = 1$, then $A(x) \geq A(x_0)$ for every x.

The area of cross-section thus has an absolute minimum at any position along the duct where $M = 1$. This shows why a supersonic tunnel requires a nozzle with a throat in order to accelerate gas in a steady flow from subsonic to supersonic speeds. To prove the lemma note that (33.3) and (33.5) imply $a^2\rho^{1-\gamma} = \gamma p \rho^{-\gamma} = \text{const}$ in homentropic flow, so that (33.8) integrates to

$$\tfrac{1}{2}u^2 + \frac{1}{\gamma - 1}a^2 = \text{const},$$

and therefore

$$u^2\left(1 + \frac{2}{\gamma - 1}M^{-2}\right) = \text{const}.$$

Thus u is a strictly monotone, increasing function of M, and it follows from (33.9) that A is also such a function of M for $M > 1$ but is a strictly monotone, decreasing function of M for $M < 1$.

34. Flow Initiation

The question of how a fluid motion starts has been avoided in the preceding chapters. The simplest, realistic form of the problem concerns a straight pipe filled with fluid and closed at one end by a piston. If the piston is pushed into the fluid, how is the motion communicated to fluid which is not close to the piston? (A similar process occurs when a valve is opened: the fluid on one side of the valve acts, roughly, like a piston on the fluid on the other side.)

On ideal fluid theory the symmetry of the problem indicates a one-dimensional motion with velocity u in the direction of the pipe axis dependent only on time t and distance x along that axis. The mass conservation equation (4.5) then implies $\partial u/\partial x = 0$, i.e., $u = u(t)$, which must be identical with the normal velocity of the piston surface at the time t, by Postulate IXb. In other words, the piston velocity is communicated instantaneously to the entire fluid, however far it may be from the piston.

The root of this paradox cannot lie in the assumption of one-dimensional motion, for if flow initiation by the radial expansion of a spherical boundary be envisaged, only a spherically symmetrical solution is at all plausible, even for a viscous fluid, and (4.5) then leads again to the implausible prediction that velocity is communicated instantaneously over arbitrary distances. The paradox must therefore stem from (4.5), which characterizes incompressible fluid motion. By contrast, compressible, inviscid fluid motion will be shown now to possess a mechanism which explains flow initiation.

It is natural to expect this process to be continuous and to take as an initial condition that the gas is at rest and uniform pressure and temperature. It follows from the equation of state that the entropy is initially uniform, and from Corollary 33.1 that the motion is homentropic. From (33.2), accordingly, grad $p = a^2$ grad ρ in (33.6).

The governing equations (33.6, 7) now take a simple form in the *acoustic approximation*, which is defined as the limit, as $\varepsilon \to 0$, of solutions

$$\rho = \rho_0(1 + \varepsilon\rho'(\mathbf{x}, t)), \qquad \mathbf{v} = \varepsilon\mathbf{v}'(\mathbf{x}, t)$$

such that ρ' and $\mathbf{v}'$ possess continuous and bounded first derivatives on the closure of the fluid domain. For such solutions (33.6, 7) reduce to

$$(34.1) \qquad \rho_0 \frac{\partial \mathbf{v}'}{\partial t} + a_0{}^2 \operatorname{grad} \rho' = O(\varepsilon), \qquad \frac{\partial \rho'}{\partial t} + \rho_0 \operatorname{div} \mathbf{v}' = O(\varepsilon),$$

where $a_0 = a(\rho_0, S_0)$. A further differentiation (if permissible) then leads to

$$\frac{\partial^2 \rho'}{\partial t^2} - a_0{}^2 \nabla^2 \rho' = \frac{\partial^2 \mathbf{v}'}{\partial t^2} - a_0{}^2 \nabla^2 \mathbf{v}' = 0,$$

the wave equation with signal speed a_0. Accordingly, the variable of state $a(\rho, S)$ is called the *speed of sound*.

In this case of one-dimensional motion suggested by the symmetry of the piston problem, $\mathbf{v}' = \{u(x, t), 0, 0\}$ and ρ' depend only on time and distance x along the pipe axis, so that (34.1) is equivalent to

$$\frac{\rho}{\rho_0} + \frac{u}{a_0} = F(x - a_0 t), \qquad \frac{\rho}{\rho_0} - \frac{u}{a_0} = G(x + a_0 t)$$

in the limit $\varepsilon \to 0$. This shows explicitly the role of $a(\rho, S)$ as the propagation speed in the wave propagation process described by (34.1). Moreover, if the gas is initially at rest and uniform pressure and temperature in $x > 0$, then $F(x) = G(x) = 1$ for $x > 0$, which implies that the gas remains at rest and at the initial pressure and temperature at least until $t = x/a_0$ for any $x > 0$; the motion is not communicated instantaneously to any fluid at a positive distance from the piston.

For the piston problem it is natural to specify the piston position $x = X(t)$ for $t \geq 0$ so that $X(0) = X'(0) = 0$ and—for consistency with the acoustic approximation—$\varepsilon^{-1} X'(t)$ is continuous and bounded. The boundary condition for the gas in $x > X(t)$ is then

(34.2) $u(X(t), t) = \varepsilon^{-1} \, dX/dt \qquad$ for $\ t > 0,$

from (3.4) or from Postulate IXb. Accordingly, $F(X(t) - a_0 t) = 1 + 2X'(t)/a_0$ for $t > 0$, and since $|X(t)| = O(\varepsilon)$, $F(z) = 1 + 2X'(-z/a_0)/a_0$ for bounded $z < 0$ in the limit $\varepsilon \to 0$. Thus

(34.3) $u(x, t) = a_0 \rho'(x, t) = 0 \qquad\qquad$ for $\ x > 0, 0 \leq t \leq x/a_0$

$$= \varepsilon^{-1} X'\left(t - \frac{x}{a_0}\right) \qquad \text{for} \ \ x > 0, t \geq x/a_0,$$

i.e., on the acoustic approximation, the gas velocity at x is that which the piston had x/a_0 seconds earlier.

The exact solution of (33.6, 7) for the piston problem will be discussed in Appendix 34 to establish the conditions under which a solution of the acoustic type exists and what degree of approximation is furnished by (34.3). This will show that the main, qualitative conclusions just drawn from the acoustic approximation apply to any one-dimensional motion initiated by a normal displacement of the boundary of a body of fluid. The situation is different for a motion initiated by a tangential displacement of the boundary; Rayleigh's problem 19.1 is then relevant and the mechanism of flow initiation is purely viscous. But its effective penetration distance is shown by Rayleigh's problem to be only $2(\nu t)^{\frac{1}{2}}$. The compressible mechanism, by contrast, has the penetration distance at. Kinetic theory [17] shows ν/a^2 to be roughly the average free time of a molecule between successive collisions.

The solution of the piston problem shows in detail how the inviscid mechanism of flow initiation acts by the propagation of a wave of velocity change, pressure change, and density change. Of course, a real pipe is not infinitely long, and the wave soon reaches its end, is reflected and travels back to the piston, where it is reflected again, and so on. If a steady flow (or an unsteady, incompressible motion) develops, it must be expected to do so as the asymptotic result of such a process, as $a_0 t/L \to \infty$, where L is a length representative of the spatial extent of the fluid domain. (In fact, the differential equations of incompressible fluid dynamics are seen to result formally from letting $a \to \infty$ in those discussed in this chapter.) In air at normal temperature and pressure the speed of sound is about 1,100 feet per second, so if $L = 10$ feet, $a_0 t/L = 100$ after about 1 second. The wave propagation process is more complicated for three-dimensional motion, but possesses the same basic nature (Huyghens' principle applies to inviscid gas motion).

Appendix 34

Simple Waves. For the discussion of the one-dimensional homentropic motion of inviscid gas it is convenient to employ an auxiliary variable of state $\omega(\rho)$, defined (up to a constant) by $\rho\, d\omega/d\rho = a(\rho, S_0)$.† Then since $\mathbf{v} = \{u(x, t), 0, 0\}$ and $\rho\,(x, t)$ depend only on time and distance x along the pipe axis, (33.6.7), reduce to

$$\frac{\partial u}{\partial t} + u \frac{\partial u}{\partial x} + a \frac{\partial \omega}{\partial x} = 0, \qquad \frac{\partial \omega}{\partial t} + u \frac{\partial \omega}{\partial x} + a \frac{\partial u}{\partial x} = 0$$

by (33.2) and (4.1). It may be noted that the context of pipe and piston is not essential to this formulation. The same equations are of interest, for instance, in astrophysics, and they are often referred to as the equations of homentropic plane waves "of finite amplitude." Addition and subtraction, and setting

(34.4) $$r = \tfrac{1}{2}(\omega + u), \qquad s = \tfrac{1}{2}(\omega - u),$$

transforms them into

(34.5) $$\frac{\partial r}{\partial t} + (u + a) \frac{\partial r}{\partial x} = 0,$$

(34.6) $$\frac{\partial s}{\partial t} + (u - a) \frac{\partial s}{\partial x} = 0.$$

These equations, due to Riemann, are a hyperbolic system, in fact, they are in characteristic form [45], and they therefore describe a process of wave propagation and interaction. Observe also that they are equivalent to the statements

(34.7) $$dr/dt = 0 \quad \text{when} \quad dx/dt = u + a,$$

(34.8) $$ds/dt = 0 \quad \text{when} \quad dx/dt = u - a.$$

† For a perfect gas $\omega = 2a/(\gamma - 1)$, by (33.3) and (33.5).

This property of r and s has led to their being called Riemann invariants. It should be noted that while (34.5, 6) depend on the differentiability of u and a, (34.7, 8) remain meaningful if they are only continuous functions. They can, in fact, be derived on that assumption from Postulates V and VI, and are therefore the appropriate generalization of Riemann's equations to continuous motion, if we do not wish to appeal to the smoothness convention.

For the discussion of (34.7, 8) it is convenient to think of the fluid domain Ω_t as a set Ω in the xt-plane. The curves in Ω defined by $dx/dt = u \pm a$ are called *Mach lines* or *characteristic lines*. For any continuous motion the Mach directions $dx/dt = u \pm a$ are continuous functions on Ω; if $P \in \Omega$, a Mach line of each family can therefore be traced from P to $\partial\Omega$ in the sense of increasing time, and similarly in the sense of decreasing time. Of course, these lines are curves defined in terms of the solution of (34.5, 6), rather than in terms of functions of x and t known a priori, and they therefore depend also on the initial and boundary conditions. To distinguish between the two families of curves, it is customary to call those on which r is constant *advancing Mach lines* (because $a > 0$, and hence the velocity $u + a$ everywhere exceeds the local gas velocity u), and those on which s is constant, *receding Mach lines*.

It is also convenient to define a *simple wave* as a subdomain on which $r \equiv$ const or $s \equiv$ const. To distinguish them, a subdomain on which only $s \equiv$ const is called an *advancing* simple wave, and one on which only $r \equiv$ const, a *receding* simple wave. A simple wave in which both $r \equiv$ const and $s \equiv$ const is called a *uniform region*, since (34.4) shows u and all variables of state there to be also independent of x and t.

In the piston problem, every point P of the fluid domain Ω is connected to the positive x-axis by a receding Mach line, because such a Mach line can be traced from P to $\partial\Omega$ in the sense of decreasing time, and it cannot intersect the piston path, by (34.8) and since $a > 0$. The initial condition is that the gas is at rest and in uniform thermodynamic state, i.e.,

$$(34.9) \qquad \left. \begin{array}{l} u = 0, \quad a = \text{const} = a_0 \\ r = s = \text{const} = s_0 = \omega_0/2 \end{array} \right\} \quad \text{at } t = 0 \quad \text{for all } x > 0.$$

By (34.8), therefore, $s \equiv s_0$ on $\Omega = \{t > 0, x > X(t)\}$, so that the fluid domain must consist of advancing simple waves and uniform subdomains.

It follows from (34.4) that the Mach line slopes $u \pm a$ in Ω depend only on r, and by (34.7) the advancing Mach lines must be straight lines. Those intersecting the positive x-axis must therefore be the lines $x - a_0 t = \text{const} > 0$, and since $r = s_0$ on these, by (34.9), the gas must remain at rest and in the initial thermodynamic state until $t = x/a_0$ at any $x > 0$.

This argument, due to Riemann, can be extended to prove uniqueness of solutions under more general circumstances [46]. It gives constructive uniqueness proofs. For instance, it can be used to prove the following Simple Wave Theorem [47]. If A and B are open subsets of a fluid domain which are adjacent in the sense that $\bar{A} \cap \bar{B}$ contains a continuous curve segment of positive length, and if the motion is uniform on A, then B contains a simple wave.

In the piston problem—for a perfect gas (i.e., $\omega = 2a/(\gamma - 1)$), for definiteness—the advancing Mach lines not intersecting the positive x-axis must intersect the piston path $x = X(t)$ (Fig. 34.1) as they are traced in the sense of decreasing time. The boundary condition there is (34.2) (with $\varepsilon = 1$, without loss of generality), and since $s \equiv s_0$ on Ω,

$$(34.10) \qquad r(X(t), t) = dX/dt + s_0$$

on the piston curve, by (34.4). Those Mach lines are therefore the lines

$$(34.11) \qquad x - X(\tau) = \left[\frac{\gamma + 1}{2} r(X(\tau), \tau) - \frac{3 - \gamma}{2} s_0 \right] (t - \tau),$$

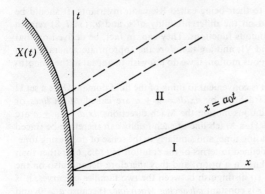

Fig. 34.1.

and $r(x, t)$ is determined in the advancing simple wave (region II, Fig. 34.1) by (34.11), (34.7), and (34.10).

Conversely, existence of the solution of the piston problem for a piston path $x = X(t) \in C^2[0, \infty)$, with $X'(0) = 0$ and $X'(t)$ monotone decreasing, may now be established by confirming the continuous differentiability of the function $r(x, t)$ just constructed. To this end, note that the direction of the straight line (34.11) is given by

$$dx/dt = \frac{\gamma + 1}{2} X'(\tau) + \text{const} = \lambda(\tau),$$

so that $d\lambda/dX' = (\gamma + 1)/2 > 0$. The monotone decrease of $X'(t)$ therefore implies that the advancing Mach lines diverge, as they are followed away from the piston path in the sense of increasing t (Fig. 34.1). It follows that the value of $|\partial r/\partial t| + |\partial r/\partial x|$ at any point P in region II (Fig. 34.1) is bounded by its value at the point Q of the piston path which lies on the same advancing Mach line as P. It therefore suffices to show that $r(X(t), t) \in C^1[0, \infty)$, and that follows from (34.10), since $X(t) \in C^2[0, \infty)$.

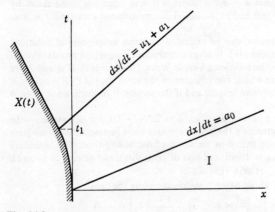

Fig. 34.2.

A special case of interest is that where the piston velocity becomes constant after a certain time, say $X'(t) = u_1$ for $t \geq t_1 > 0$. The advancing simple wave is then only the subset of region II between the wave front $x = a_0 t$ and the advancing Mach line through $(X(t_1), t_1)$ (Fig. 34.2). Since $r(X(t), t) = u_1 + s_0 = $ const for $t \geq t_1$, the remainder of region II is a uniform region, in which

$$(34.12) \qquad u \equiv u_1, \qquad a \equiv \frac{\gamma - 1}{2}(u_1 + 2s_0) = a_1$$

by (34.4). Now let $t_1 \to 0$, keeping $X'(t_1) = u_1$ fixed, to obtain the limiting case of a piston suddenly brought from rest (at $t < 0$) to the velocity $u_1 < 0$ (at $t > 0$). Since the slopes of the advancing Mach lines depend only on the values of r on the piston curve, the advancing simple wave does not disappear as $t_1 \to 0$ (Fig. 34.3). Its limit is called a *centered simple wave*. (The limiting solution is continuous, in fact piecewise continuously differentiable on the fluid domain, even if not on its closure.)

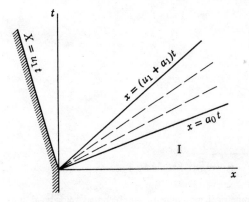

Fig. 34.3.

Comparison with the exact solution defined by (34.7) and (34.10, 11) shows the acoustic approximation (34.3) to give the velocity and density (but not their derivatives) with an error of order ε only, if no other limit process is relevant. But, for instance, the acoustic approximation fails for a centered wave ($t_1 \to 0$). Similarly, it predicts $u = $ const on any line $x - a_0 t = $ const < 0, while the exact solution of Riemann's equations shows $u \to 0$ as $t \to \infty$ on any such line. It also fails to reveal the difficulties to be discussed now.

The significance of the monotone decrease of piston velocity for the existence of the solution may be illuminated further by consideration of the case of a strictly increasing piston velocity. The advancing Mach lines in region II (Fig. 34.1) now converge with increasing time, and since they are straight they must form an envelope after a certain time. Moreover, each of these Mach lines carries its own value of r, and if $X'(t)$ is strictly monotone, two distinct, advancing Mach lines cannot carry the same value of r, by (34.10). It follows that $(\partial r/\partial t)^{-1} = (\partial r/\partial x)^{-1} = 0$ at the envelope, and hence the solution fails to be differentiable there. In fact, the fluid acceleration Du/Dt is singular at the envelope, because $Du/Dt = -a\,\partial r/\partial x$ in an advancing simple wave by (3.1), (34.4), and (34.5).

Such an envelope is called a limit line because the singularity heralds a more than local breakdown of the solution [46]. This is most easily seen at the instance of the centered simple wave corresponding to a positive piston velocity u_1. Let S denote the triangular, centered simple wave region (Fig. 34.3) between the advancing Mach lines $x = a_0 t$ and $x = (u_1 + a_1)t$ bounding the respective regions of uniform motion. For $u_1 > 0$, (34.12) gives $u_1 + a_1 > a_0$ and it follows that three distinct, advancing Mach lines pass through each point of S (Fig. 34.4); two of these Mach lines belong to the respective regions of uniform motion and carry the corresponding values r_0 and $r_1 \neq r_0$ of r, and the third Mach line belongs to the "centered wave" and carries an intermediate value of r. Accordingly, the gas velocity and the speed of sound (and hence also the pressure, density, and temperature) are triple-valued in S—that is, a centered simple wave solution does not exist for positive piston velocity (but see Appendix 36).

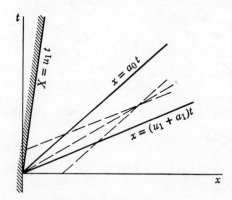

Fig. 34.4.

Problem 34.1. In spherically symmetrical motion with purely radial velocity u, the equation of continuity, (4.5), is

$$x^2 \, \partial \rho / \partial t + \partial(x^2 \rho u)/\partial x = 0$$

if x denotes the radius. For homentropic, inviscid gas motion with this symmetry, derive the characteristic equations for Riemann's $r(x, t)$ and $s(x, t)$. What simple wave motions ($r \equiv$ const or $s \equiv$ const) and what steady motions are possible?

35. Conservation of Energy

The First Law of Thermodynamics postulates that any thermodynamic system possesses a variable of state, the *internal energy E* per unit mass, such that, in any transition from one equilibrium state to another, the difference between the internal energy change and the work done by external forces on unit mass of the system must be supplied to the system in the form of heat.

A transition is called *adiabatic* if the total heat supplied in it is zero (i.e., as much heat is added as is taken away). Moreover, a system is called *isolated* if it is incapable of exchanging heat with its surroundings, but since it is natural to regard a fluid as a physical system rather than a purely thermo-dynamical one, it will be more convenient here to use the term *strictly adiabatic* for a transition involving no heat addition at any stage.

For a fluid in motion it is plausible (and consistent with kinetic theory) to interpret the internal energy as the sum of the kinetic energy $|\mathbf{v}|^2/2$ per unit mass and the same variable of state E which measures the internal energy in the fluid at rest; and this will be regarded as part of the basic gas-dynamical postulate. The nature of the work done by external forces is settled by the momentum principle (Section 10), which defines those forces. The physical processes by which heat can be added to a body of fluid are conduction and radiation, and the latter will be ignored in this chapter, as in much of the gas dynamic literature. The conduction of heat is described by a *heat flux* vector field $\mathbf{q}$ (Appendix 18). The First Law of thermodynamics, or principle of energy conservation, for a fluid in local thermodynamic equilibrium is therefore

$$(35.1) \quad \frac{D}{Dt}\int_{\Omega_t}(\tfrac{1}{2}|\mathbf{v}|^2 + E)\rho\, dV = \int_{\Omega_t}\rho\mathbf{f}\cdot\mathbf{v}\, dV + \int_{\partial\Omega_t}p_{ij}n_jv_i\, dS - \int_{\partial\Omega_t}\mathbf{q}\cdot\mathbf{n}\, dS$$

for every fluid domain Ω_t with regular boundary surface $\partial\Omega_t$.† The first two terms on the right-hand side represent the rate at which the external forces do work on the body of fluid occupying Ω_t; the contribution of the body force field $\mathbf{f}$ has been included for reference in Section 37, but for simplicity will be neglected until then. As noted in Appendix 18, the heat flux vector $\mathbf{q}$ is given by Fourier's law,

$$(35.2) \quad \mathbf{q} = -\lambda\, \text{grad}\, T,$$

where the coefficient λ of heat conduction is a function of the local state only.

Following general usage, we now restore the name internal energy to the variable of state E. From a physical point of view it may be divided in two additive parts, the communicable and the inert energy, of which the latter represents the contribution of all those internal mechanisms of the substance which take no part in the processes considered and therefore conserve their energy content. Since the First Law concerns only differences in E, it suffices to define this variable as the communicable internal energy, and that is, in fact, its thermodynamic meaning; specification of *this* internal energy is part of the definition of a thermodynamical system. Now the consideration of dissociation [48] or chemical reactions is beyond the scope of this book, and the energy of atomic and molecular bonds may thus be regarded as inert. Similarly, condensation and other phase transitions are not of direct interest

† The notation is defined in Sections 2 and 3; no confusion will arise from the use of dS for the surface element and S for the entropy, since the latter will not be used as a variable of integration.

in what follows. It will suffice to consider only that energy which is changed in a gas by moderate changes in pressure and temperature, and which kinetic theory interprets as the kinetic energy of the mechanical degrees of freedom of the random part of the molecular motion. For a perfect gas (of constant specific heat, as always in this chapter) this is [49]

$$(35.3) \qquad E = c_v T,$$

apart from an additive constant.

The First Law, like the other conservation principles, is a Lagrangian statement (Section 2), and for applications it is convenient to translate it into the Eulerian framework by the help of Corollary 3.1. For steady flow this gives

$$\frac{D}{Dt} \int_{\Omega_t} (\tfrac{1}{2}|\mathbf{v}|^2 + E)\rho \, dV = \int_{\partial\Omega_t} (\tfrac{1}{2}|\mathbf{v}|^2 + E)\rho \mathbf{v}\cdot\mathbf{n} \, dS,$$

so that the First Law (without body force and radiation) takes the form

$$(35.4) \qquad \int_{\partial\Omega} [(\tfrac{1}{2}|\mathbf{v}|^2 + E)\rho\mathbf{v}\cdot\mathbf{n} - p_{ij}n_jv_i + \mathbf{q}\cdot\mathbf{n}] \, dS = 0$$

for every fixed domain Ω with regular boundary occupied by gas in steady flow.

In the next two sections, the First Law will be needed specifically for steady, one-dimensional flow in the x-direction. It is then convenient to apply (35.4) to the part $x_1 < x < x_2$ of the interior of a cylinder of unit cross-sectional area with generators parallel to the direction of flow. By symmetry, only the subsets of $\partial\Omega$ on which $x = x_1$ or $x = x_2$ then contribute to the integral, and the integrand of (35.4) must therefore take the same value at x_2 as at x_1. But those values of x are still arbitrary, and therefore

$$(35.5) \qquad (\tfrac{1}{2}u^2 + E)\rho u - p_{11}u + q_1$$

is independent of x. The first term represents the rate at which the flow carries energy in the direction of increasing x across unit area of a plane $x = $ const; the second term represents the rate at which the stress on that area does work on the gas at smaller values of x; and the last term represents the heat conducted to that gas across that area.

The clearest application of this result is to the case where the motion and thermodynamic state are uniform in a neighborhood of x_1 and again in a neighborhood of x_2. Then $q_1 = 0$ on those neighborhoods, by (35.2), and the body of gas occupying the domain Ω between x_1 and x_2 at time t is in strictly adiabatic motion during a time interval including t. Conservation of mass, moreover, implies that ρu is independent of x, and by Postulate VII, $p_{11} = -p$ both at $x = x_1$ and at $x = x_2$, so if subscripts are used to distinguish values at x_1 and x_2, (35.5) reduces to

$$(35.6) \qquad \tfrac{1}{2}u_2{}^2 + E_2 + p_2/\rho_2 = \tfrac{1}{2}u_1{}^2 + E_1 + p_1/\rho_1.$$

This suggests the definition of a further variable of state,

(35.7) $$I = E + p/\rho,$$

called *enthalpy* per unit mass, and a quantity

(35.8) $$I_0 = I + |\mathbf{v}|^2/2,$$

called *stagnation enthalpy*, in terms of which (35.6) becomes simply

(35.9) $$I_0(x_2) = I_0(x_1).$$

Stagnation enthalpy is, in fact, the most natural, local measure of total energy per unit mass in a compressible fluid, and plays as prominent a role in gas dynamics as Bernoulli's total pressure (Section 12) plays in incompressible fluid dynamics.

A more sophisticated application of the First Law is to the case where the flow at x_1 and x_2, though not uniform, is such that an inviscid limit is approached *there*. That depends on the local Reynolds number, a plausible definition of which is $R(x) = u^2/(\nu\, du/dx)$. However, Postulate IXa gives

(35.10) $$p_{11} = -p + (\mu' + 2\mu)\, du/dx,$$

so that $\nu = \mu/\rho$ should be replaced by $(\mu + \mu'/2)/\rho$. Moreover, since heat conduction and viscous momentum transport are closely related processes, small viscosity of a gas can be described consistently only by a parameter playing a role analogous to the Reynolds number also for heat conduction. These requirements are met by the choice

$$R(x) = \min\left[\frac{\rho u^2}{(\mu' + 2\mu)\, du/dx}, \frac{\rho u^3}{\lambda\, dT/dx}\right]$$

for the local Reynolds number. Then as $R(x_2) \to \infty$ and $R(x_1) \to \infty$, the heat conducted across unit area of the planes $x = x_2$ and $x = x_1$ tends to zero, by (35.2), and the flow of the body of gas between these planes tends to a strictly adiabatic one. By (35.2) and (35.10), moreover, the First Law (35.5) implies again (35.6) and (35.9) in that limit. It may also be noted that there are several ways in which an approach to a limit such as $R(x_1) \to \infty$ may be realized; one way, of which the physical relevance will become apparent in Section 37, is by means of a sequence of values of x_1 such that the corresponding values of $R(x_1)$ increase monotonely beyond bounds.

A simpler application is to an inviscid fluid (Section 13), which by definition is incapable of conducting heat and hence is always in strictly adiabatic motion. Since also $p_{ij} = -p\, \delta_{ij}$, the First Law then implies $I_0(x) = $ const for $x_1 \le x \le x_2$.

Appendix 35

Second Law. Thermodynamics distinguishes particular (limiting) transitions between states of a system which can be *reversed* so that both the system and its surroundings return ultimately to their respective initial states. It is a principle of thermodynamics [50] that such reversible transitions must proceed through a continuous succession of thermal equilibrium states depending differentiably on a suitable parameter τ. The Second Law postulates the existence of variables of state T and S such that, during any reversible transition of a system

$$(35.11) \qquad DS/D\tau = T^{-1} \, dQ/d\tau,$$

where $dQ/d\tau$ is the rate (with respect to the parameter) at which heat is added to the system during this transition, per unit mass. (The differentiation symbols in (35.11) are meant to emphasize that one represents the Lagrangian notion of rate of change of a property of the system while the other represents the rate of an external action upon the system.) The Second Law further postulates that the entropy change in an arbitrary transition between two states of a system cannot be smaller than $\int T^{-1} \, dQ$, where Q refers to the actual heat addition occurring during that transition and T is the temperature at which this heat addition takes place.

It is an immediate corollary that the entropy of an isolated system cannot decrease.

For a gas at rest thermodynamics defines the rate of heat addition in a reversible transition, consistently with the First Law (35.1), as

$$(35.12) \qquad dQ/d\tau = DE/D\tau + p \, D(1/\rho)/D\tau,$$

because the work done by the pressure can be stored in a spring and then restored to the gas to reverse the transition if the changes proceed sufficiently slowly. With the reversible rate of heat addition thus defined, the Second Law actually defines the entropy (except for an additive constant). The earlier definition (33.3) for the perfect gas is consistent with this, since it is equivalent to (35.11), by (33.1), (35.3), and (35.12).

Thermodynamics proposes all this—and the notation strongly reflects that—in the first place only for systems in thermal equilibrium, and the quantities discussed are well-defined. It is a simple step to extend the Laws to any system Σ which is the union of a finite number of such subsystems. The mutual heat exchanges between the subsystems do not then contribute to the addition of heat to Σ. To define the heat added to Σ, and also the lower bound to the entropy change of Σ, it therefore suffices to consider those subsystems which exchange heat directly with the surroundings of Σ. For such purposes it is then immaterial whether the other subsystems are in thermal equilibrium.

The application of the Second Law to fluid motion depends on the principle that it applies also to any limit of such unions of subsystems for which the quantities under discussion are well-defined. To illustrate this limit principle, consider a body of gas occupying a fluid domain Ω_t with regular boundary surface $\partial\Omega_t$. In the absence of radiation the limit principle then implies that the rate of heat addition to the body of gas is given by the last integral of (35.1) if the heat flux vector is defined on $\partial\Omega_t$. If, in addition, the boundary values of the temperature are defined, then the lower bound to the rate of change of the entropy of the body of gas is

$$- \int_{\partial\Omega_t} T^{-1}\mathbf{q}\cdot\mathbf{n} \, dS.$$

It follows that the entropy of such a body of gas cannot decrease in a strictly adiabatic motion.

Theorem 33.1 now follows from the First and Second Laws. In the absence of a body force field, work can be done on an inviscid gas only by the pressure, and such work is reversible. In the absence of radiation, moreover, any motion of inviscid gas must be strictly adiabatic, because such gas is devoid of heat conduction. By the First Law, any transition must therefore be reversible, and by (35.11) the entropy cannot change.

The theorem will be elucidated further in Section 38 by a calculation of the entropy change caused by the irreversible work of the stresses and by heat conduction. This will also show that the absence of body forces is not a necessary condition for the theorem.

36. Shock Relations

For an incompressible fluid, strictly one-dimensional, steady flow is trivial, because (4.5) implies that the velocity $u(x)$ = const. For inviscid gas in continuous, homentropic motion, such flow is similarly trivial, by (33.7, 8). To see that it is not generally trivial for a gas, we now apply all the conservation principles to the domain Ω which is the part $x_1 < x < x_2$ of the interior of a cylinder of unit cross-section with generators parallel to the direction of flow, which is taken to be the x-direction. For definiteness we begin by assuming that the flow and state are uniform in neighborhoods of x_1 and x_2, respectively.

Conservation of mass implies

(36.1)
$$\rho_2 u_2 = \rho_1 u_1,$$

by (4.2), where the subscripts again denote the values assumed at x_1 and x_2, respectively. Conservation of momentum implies

(36.2)
$$p_2 + \rho_2 u_2{}^2 = p_1 + \rho_1 u_1{}^2,$$

by (10.4) and Postulate VII. Conservation of energy has been shown to imply (35.9), i.e.,

(36.3)
$$I_2 + \tfrac{1}{2}u_2{}^2 = I_1 + \tfrac{1}{2}u_1{}^2,$$

and this may be written

(36.4)
$$\tfrac{1}{2}u_2{}^2 + \frac{\gamma}{\gamma - 1}\frac{p_2}{\rho_2} = \tfrac{1}{2}u_1{}^2 + \frac{\gamma}{\gamma - 1}\frac{p_1}{\rho_1}$$

for a perfect gas, since then $I = c_p T = \gamma p/[(\gamma - 1)\rho]$ by (35.7), (35.3), (33.1), and (33.4).

Suppose that p_1, ρ_1, and u_1 are given; then (36.1, 2, 4) are a system of three algebraic equations for p_2, ρ_2, and u_2. Since they are not linear in these unknowns, there may be more than one solution. To test this, exclude the trivial solution by assuming $\rho_2 \neq \rho_1$, $u_2 \neq u_1$, and eliminate u_2 from (36.1) and (36.2) to obtain

(36.5)
$$\rho_1/\rho_2 = 1 + (\gamma M_1{}^2)^{-1}(1 - p_2/p_1)$$
where

$$M_1{}^2 = u_1{}^2/a_1{}^2 = \gamma p_1/(\rho_1 u_1{}^2),$$

by (33.6). Then eliminate u_2 from (36.1) and (36.4), and ρ_1/ρ_2 from the result, with the help of (36.5), to get

$$(36.6) \qquad \frac{p_2}{p_1} = 1 + \frac{2\gamma}{\gamma + 1}(M_1{}^2 - 1).$$

This shows that the conservation principles for the steady flow from x_1 to x_2 do have a nontrivial solution if $M_1{}^2 \neq 1$. A flow corresponding to it is called a *shock*. Substitution of (36.6) into (36.5) gives the density ratio as

$$(36.7) \qquad \frac{\rho_2}{\rho_1} = \frac{(\gamma + 1)M_1{}^2}{2 + (\gamma - 1)M_1{}^2},$$

and the velocity ratio follows from (36.1).

A similar, though less explicit, calculation can be carried out also for a nonperfect gas. (From (36.1, 2), $(\rho_1{}^{-1} + \rho_2{}^{-1})(p_2 - p_1) = u_1{}^2 - u_2{}^2$, and (36.3) and (35.7) now give the Hugoniot Relation

$$2(E_2 - E_1) = (p_2 + p_1)(\rho_1{}^{-1} - \rho_2{}^{-1})$$

between the thermodynamic states on the two sides of the shock. It can be used [47] to show that the qualitative conclusions to be drawn below from (36.6, 7) apply to a very general class of gases.

It is worth emphasizing that the argument just presented is an exact mathematical analysis of the type encountered in Section 11. It is a direct application of the global conservation principles of (classical) physics to the fluid domain $x_1 < x < x_2$. Under the assumptions stated those principles imply rigorously that the uniform states at x_1 and x_2, if not identical, must satisfy the *shock relations* (36.1–3)

In fact, the analysis of this section is valid under even much weaker assumptions! Consider the body of gas which occupies the domain Ω at an arbitrarily chosen time t and suppose that it is in one-dimensional motion. If it be assumed that the gas possesses an energy e per unit mass such that the energy per unit volume is integrable, then the total energy of this body of gas is

$$\int_{x(a_1, t)}^{x(a_2, t)} e\rho \, dx,$$

where a_i ($i = 1, 2$) denotes the Lagrangian coordinate (Section 2) coincident with x_i at t. It is now sufficient to assume that the motion is steady, that $e\rho$ is continuous on neighborhoods of x_1 and x_2, respectively, and that Postulate IV applies there, to obtain $e_2\rho_2u_2 - e_1\rho_1u_1$ for the rate of change of the total energy. Moreover, if the gas is in local thermodynamic equilibrium at x_1 and x_2, then $e_i = \frac{1}{2}u_i{}^2 + E_i$, where E is the variable of state representing the internal energy. Finally, if the flow is uniform on those neighborhoods and radiation is absent, then there can be no heat flux in those neighborhoods, and Postulate VII and the basic statement of the First Law for the body of gas as a general thermodynamic system imply that $pu + (\tfrac{1}{2}u^2 + E)\rho u = \rho u I_0$ must take the same value on both neighborhoods.

Similarly, if the gas has integrable mass and linear momentum per unit volume, if the motion is one-dimensional and steady, and if the gas behaves like a continuum fluid in neighborhoods of x_1 and x_2, respectively, its rates of change of mass and x-momentum are respectively $\rho_2 u_2 - \rho_1 u_1$ and $\rho_2 u_2^2 - \rho_1 u_1^2$. Thus if the flow is uniform on those neighborhoods and if the influence of body forces is negligible, then Postulate VII and the basic statements of mass and momentum conservation for the body of gas as a general physical system imply (36.1) and (36.2). The validity of the shock relations therefore extends far beyond the realm of gas dynamics!

The nature of the transition represented by the nontrivial solution is greatly illuminated by the Second Law of thermodynamics (Appendix 35), which implies that the entropy of a body of gas—considered again as a general thermodynamic system—cannot decrease in a strictly adiabatic motion. Under the assumptions just reviewed, the body of gas occupying Ω at time t is in such a motion, because the uniformity of the state near $x = x_1$ and $x = x_2$ precludes heat conduction across those planes. Since the flow is one-dimensional and steady, the rate of change of the entropy of this body of gas is $S_2 \rho_2 u_2 - S_1 \rho_1 u_1$, and therefore

$$(36.8) \qquad S_2 \rho_2 u_2 - S_1 \rho_1 u_1 \geq 0.$$

With the abbreviation

$$\sigma = M_1^2 - 1,$$

(32.4), (36.6), and (36.7) show the entropy difference for a perfect gas to be given by

$$\begin{aligned}
\frac{S_2 - S_1}{c_v} &= \log \frac{p_2}{p_1} - \gamma \log \frac{\rho_2}{\rho_1} \\
&= \log \left(\frac{2\gamma}{\gamma + 1} \sigma + 1 \right) + \gamma \log \left[1 + \frac{\gamma - 1}{\gamma + 1} \sigma \right] \\
&\quad - \gamma \log (1 + \sigma) \\
&= f(\sigma),
\end{aligned}$$

say. Now $f(0) = 0$ and

$$\frac{df}{d\sigma} = \frac{2\gamma(\gamma - 1)\sigma^2}{(1 + \sigma)[\gamma + 1 + (\gamma - 1)\sigma](\gamma + 1 + 2\gamma\sigma)} > 0$$

because $\gamma > 1$ (by (33.4)) and $1 + \sigma = M_1^2 > 0$ and $\gamma + 1 + (\gamma - 1)\sigma = 2 + (\gamma - 1)M_1^2 > 0$ and $\gamma + 1 + 2\gamma\sigma = (\gamma + 1)p_2/p_1 > 0$. Hence, $f(\sigma)$, and therefore also $S_2 - S_1$, has the same sign as σ, i.e., as $M_1^2 - 1$.

It has already been noted that conservation of mass in one-dimensional, steady flow implies $\rho u = $ const, and if the uninteresting case of gas at rest is excluded ρu is therefore of definite sign, and no generality is lost in choosing the sense of x increasing so that $u > 0$ and hence also $M_1 > 0$. The inequality (36.8) therefore implies that the nontrivial transition corresponding to the shock relations can be possible only if $M_1 = u_1/a_1 > 1$. Since a is the speed of sound (Section 34), such a flow is called supersonic. It follows from (36.6) and (36.7) that $p_2 > p_1$, $\rho_2 > \rho_1$, so that the transition must be a compression; by (36.1), therefore, $u_2 < u_1$, so that it must be a deceleration. It is not difficult, moreover, to deduce from (36.6), (36.7), (33.1), and (33.4) that $T_2 > T_1$, so that the transition must heat the gas, and thence, from (33.5), that $M_2 < 1$, so that the shock changes the flow from a supersonic one to a subsonic one.

In the absence of neighborhoods of x_1 and x_2 on which the state is uniform the same conclusions still apply in the limit, as $R(x_1) \to \infty$ and $R(x_2) \to \infty$, to a gas which is Newtonian on such neighborhoods. This is because (35.9) then applies (Section 35), because the Momentum Principle and (35.10) then imply (36.2) and because the motion on Ω then tends to a strictly adiabatic one. And of course, the same conclusions apply a fortiori to inviscid gas.

In view of the generality of the considerations which lead to them, these conclusions establish a remarkable amount of detailed information on the nature of the shock transition. To appreciate them fully, however, it is also necessary to note that there are aspects of this process on which they shed almost no light at all.

For an inviscid gas the argument imposes no conditions on x_1 or x_2. Theorem 33.1, (36.1), and (36.8), moreover, show that a shock transition cannot be continuous for an inviscid gas. The obvious conclusion from this, frequently drawn or implied in the literature, seems to be that a shock transition is a discontinuity of inviscid gas motion. But such a conclusion is not justified by anything that has *so far* been said in this chapter. On the contrary, if a shock be a discontinuous transition, then all of continuum fluid dynamics and equilibrium thermodynamics cannot add to the preceding results anything except that this transition cannot take place *at* x_1 or *at* x_2.

On the other hand, if the transition be continuous and describable within the framework of gas dynamics, then Theorem 33.1, (36.1), and (36.8) imply the further result that the mechanism of transition must depend on viscosity or heat conduction. Beyond that the discussion of this section has not revealed any information on the internal structure and mechanism of the shock process, any more than on the thickness $x_2 - x_1$ of the shock zone. Without such information it cannot be concluded that a shock of the type discussed exists; there are physical systems which admit an analogous analysis of the conservation laws but do not possess a mechanism admitting steady transitions.

Problem 36.1. For motion of perfect gas at uniform and constant stagnation enthalpy, show that there is a definite "sonic" value a_*, dependent only on I_0 and γ, which the velocity magnitude must assume at any point where $|\mathbf{v}| = a$.

This defines the sonic speed a_*, in terms of the local value of I_0, as a local flow property of any perfect gas motion. Show that the shock relations imply Prandtl's relation $u_1 u_2 = a_*^2$, where a_* is the value of the sonic speed at $x = x_1$ and $x = x_2$.

Problem 36.2. The term "strong shock" is often used to denote the limit of the shock relations as $M_1 \to \infty$; this limit is of interest in connection with hypersonics [51], strong explosions, and astrophysics. Show that ρ_2/ρ_1, $p_2/(\rho_1 u_1^2)$, $(M_1^2 - 1)^{-1} T_2/T_1$, $(M_1^2 - 1)^{-1} \exp[(S_2 - S_1)/c_v]$, and M_2 then tend to limits for a perfect gas, and find these limits in terms of γ.

Problem 36.3. The term "weak shock" denotes the limit of the shock relations as $M_1 \downarrow 1$. With $\sigma = M_1^2 - 1$, show that $(p_2 - p)/(\sigma p_1)$, $(\rho_2 - \rho_1)/(\sigma \rho_1)$, $(T_2 - T_1)/(\sigma T_1)$, and $(S_2 - S_1)/(\sigma^3 c_v)$ then tend to limits. (Since the entropy rise is a measure of inefficiency for adiabatic processes, the result that it is only $O(\sigma^3)$ for weak shocks is of great practical significance for supersonic flight.) Compute the limit of $S_2 - (S_1)/(\sigma^3 c_v)$ for a perfect diatomic gas, such as air, for which $\gamma = \frac{7}{5}$ under normal conditions.

Appendix 36

Moving Shocks. The shock relations may be extended immediately to the case of a shock zone moving with constant velocity by appeal to the Galilean invariance of classical mechanics and thermodynamics. In particular, the observer who made the analysis of Section 36 may be regarded as one who actually moves in the direction of x increasing with constant velocity U. The observer now regarded as stationary then observes the gas velocity $v = U + u$. On the other hand, the basic gas dynamical postulate (Section 33) asserts that the definition of thermodynamic state is the same for the gas in motion and at rest, and the two observers therefore see the same values of p, ρ, and S at corresponding points. The shock relations (36.1–3) and (36.8) therefore determine p_2, ρ_2, and $v_2 = U + u_2$ if p_1, ρ_1, $v_1 = U + u_1$, and U are known. It follows—and this is more often of practical interest—that they determine p_2, ρ_2, and U if p_1, ρ_1, v_1, and v_2 are known.

In particular, if $U = -u_1$, the gas is at rest with respect to the stationary observer at $\xi_1 = x_1 + Ut$, where ξ denotes the distance in the direction of motion (still taken as that of increasing x) for the stationary observer. And since (36.1–3) and (36.8) have been shown to imply $u_2 < u_1$, it follows that $v_2 < 0$. This moving shock is therefore a transition of the gas from rest and thermal equilibrium at pressure p_1 and density ρ_1 to uniform motion with $v_2 < 0$ and $p_2 > p_1$, $\rho_2 > \rho_1$ at $\xi \geq \xi_2 = x_2 + Ut$ (Fig. 36.1). Observe

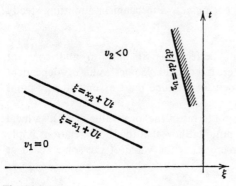

Fig. 36.1.

that the condition $M_1 = u_1/a_1 > 1$ deduced in Section 36 from the Second Law now implies $|U|/a_1 > 1$. The shock transition must therefore propagate into the gas at rest at a speed greater than the speed of sound of that gas.

Consistently with Postulate IXb, any particle path $d\xi/dt = v$ represents a possible piston path (Fig. 36.1), and comparison with Figs. 34.1 and 34.4 suggests that our present result may concern a solution of the flow initiation problem for a piston *pushed into* a gas—for which the inviscid, homentropic model of Sections 33 and 34 was shown in Appendix 34 to possess a solution for a short time at most. Indeed, in the limit $(x_2 - x_1)/(a_1 t) \to 0$ (Fig. 36.2) the shock relations are seen to furnish a single-valued but discontinuous solution for the problem of Fig. 34.4.

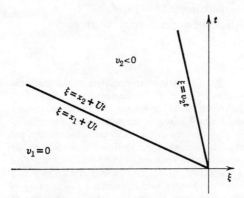

Fig. 36.2.

This result, in combination with that of Section 34, also furnishes an explanation of flow initiation in a *shock tube*, at any rate in the limit $(x_2 - x_1)/(a_0 t) \to 0$. A shock tube is a long, straight pipe initially divided in two parts by a membrane. Both parts are filled with gas, which is permitted to settle to rest and a uniform state so that its temperature becomes uniform throughout the tube, but the pressures in the two parts are arranged to be different. For definiteness suppose that the same perfect gas is used on both sides of the membrane, the position of which is taken to be $x = 0$, and that the pressure p_1 of the gas in $x < 0$ is greater than the pressure p_0 of the gas in $x > 0$. By (33.1), the density ρ_1 of the gas on the left is then also greater than that, ρ_0, of the gas on the right ($x > 0$).

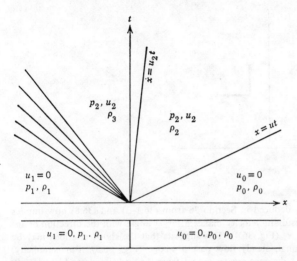

Fig. 36.3.

At $t = 0$ the membrane is made to disintegrate. On the assumption that the subsequent motion is one-dimensional, it may be described in terms of a centered simple wave of expansion (Appendix 34) propagating to the left (Fig. 36.3) and [the limit $(x_2 - x_1)/(a_0 t) \to 0$ of] a shock propagating to the right. The shock speed U is determined by the condition that both processes bring the gas between them to the same uniform velocity u_2 and pressure p_2. The simple wave solution and the shock relations imply that $u_2 > 0$ and $p_1 > p_2 > p_0$.

The line $x = u_2 t$ is the path of the interface between the left gas and the right gas, and a difference in density across this interface must be anticipated. In the simple, limiting motion just described, the interface is a discontinuity surface (Section 9), which is a one-dimensional analog of a vortex sheet. In a viscous gas such a discontinuity surface cannot persist, but, like a vortex sheet (Problem 19.1), it diffuses only slowly in a gas of small viscosity. Similarly, for consistency with Postulate IXb, a boundary layer must form on the pipe walls, but at any value of x its growth can begin only when the motion has penetrated to that value of x, and in a gas of small viscosity it grows only slowly and exerts an appreciable influence on the motion near the pipe axis only after a relatively long time.

Oblique Shocks. A similar extension of the shock relations by Galilean invariance is possible in steady flow by adoption of the point of view of an observer moving with constant velocity V in a direction normal to that of increasing x. That observer sees a steady flow but not a one-dimensional one. In addition to an x-component of velocity

$v_n = u$ normal to the shock zone, he also see a y-component $v_t = -V$ tangential to the shock zone. The shock zone is therefore inclined obliquely to his resultant velocities at x_1 and x_2, respectively (Fig. 36.4).

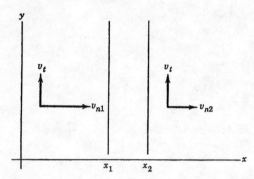

Fig. 36.4.

The condition $u_2 < u_1$ deduced in Section 36 from (36.1–3) and (36.8) now implies $v_{n2} < v_{n1}$, so that the resultant velocity makes a larger angle with the direction of increasing x at x_2 than at x_1 (Fig. 36.4). This suggests that the shock relations may be relevant to two-dimensional, steady flow past a wedge (Fig. 36.5). For simplicity, consider only the flow at a distance greater than L from the wedge apex and let $\kappa L \to 0$, where κ is the curvature of the wedge tip (Fig. 36.5). If L is the observation scale, the

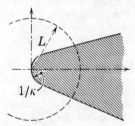

Fig. 36.5.

wedge then tends to an ideal, sharp-edged one (Fig. 36.6). Suppose that it is consistent to take also the limit $(x_2 - x_1)/L \to 0$ for the shock transition. The shock relations then describe a discontinuous transition from a uniform flow in the x-direction to a uniform flow parallel to the flank of the wedge [and hence satisfying the matching condition (3.4) on the wedge surface], provided that values of v_t can be found for which the resultant velocity at x_2 makes with that at x_1 (Fig. 36.4) the angles θ_1 and θ_2 (Fig. 36.6), respectively. These values of v_t will also determine the respective inclinations of the shock fronts to the direction of increasing x.

It should be recalled that the Second Law was shown in Section 36 to imply $u_1 > a_1$, and accordingly $M_1 \equiv |v_1|/a_1 > 1$ in Fig. 36.4. The shock relations therefore give the

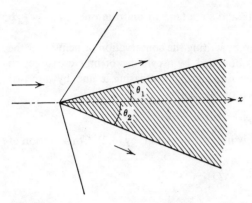

Fig. 36.6.

limiting solution for steady flow past a wedge just discussed only if the uniform flow ahead of the wedge is supersonic. But the condition $u_2 = v_{n2} < u_1 = v_{n1}$ does not imply $M_2 = |v_2|/a_2 < 1$ when $|v_t| > 0$, so that an oblique shock need not be a transition to subsonic flow. On the other hand, closer scrutiny of the shock relations shows that the value of M_1 determines an upper bound for the value of max (θ_1, θ_2) for which appropriate values of v_t can be found. For given M_1 the simple limiting solution just discussed can therefore exist only if the wedge is sufficiently slender. Moreover, scrutiny of the shock relations shows also that, if a pair of appropriate values of v_t can be found for a given triplet $(M_1, \theta_1, \theta_2)$, then a second such pair can generally also be found! The shocks corresponding to both can be observed in a wind tunnel, but fortunately only those corresponding to the smaller entropy increase are relevant to flight.

For more detail on all these maters, and also for extensions of the conservation law analysis to transitions involving chemical reactions, the reader may be referred, in the first place, to [47].

Figure 23.1 shows a steady flow past a tetrahedral body. The incident flow (to the left of the light and dark traces) is very close to a uniform one of Mach number $M_1 = 4$. On the scale L of the photograph, κL and $(x_2 - x_1)/L$ are small numbers. Since the body is three-dimensional, so is the flow, and the shock is conical, but this cone incorporates three plane faces. The straight trace sloping from the tip of the body to the upper right-hand corner of the photograph is the schlieren image of that plane face of the shock to which the light rays are (near-) tangent. The straight trace sloping toward the lower right-hand corner is the image of the lowest generator of the shock cone, where the light rays are again tangent to that cone.

37. Shock Structure

The shock relations are greatly illuminated by Becker's proof that the conservation principles possess, for a perfect, monatomic, Newtonian gas of Prandtl number $\sigma = \mu c_p/\lambda = \frac{3}{4}$, and in the absence of body force and radiation, a smooth, steady, one-dimensional solution describing a shock transition in which the velocity and state tend to uniform ones as $x \to -\infty$ and again as $x \to +\infty$.

To derive this solution, begin by writing the conservation principles for the fixed fluid domain Ω which is bounded by a cylinder of unit cross-section with generators parallel to the direction of increasing x and by *any* two planes $x = \text{const.}$ Conservation of mass gives

$$(37.1) \qquad \rho u = \text{const} = m$$

for steady, one-dimensional flow in the x-direction, by (4.2). Conservation of momentum gives

$$\rho u^2 - p_{11} = \text{const,}$$

by (10.4), and thus

$$(37.2) \qquad mu + p - (\mu' + 2\mu)\, du/dx = \text{const,}$$

by Postulate IXa (Section 17). The First Law has already been shown to give (35.5) or, by (37.1) and (35.7, 8),

$$(37.3) \qquad mI_0 - (\mu' + 2\mu)u \frac{du}{dx} - \frac{\mu c_p}{\sigma} \frac{dT}{dx} = \text{const.}$$

(It may be noted that only $(u, T) \in C^1(-\infty, \infty)$ is assumed.) For a perfect, monatomic gas, kinetic theory [19, Chap. 16] shows $\mu'/\mu + \frac{2}{3}$ to be proportional to the square of the ratio of the volume occupied by the molecules themselves to the volume of the gas, and therefore, in the limit corresponding to the Newtonian constitutive equation,

$$(37.4) \qquad \mu' + 2\mu/3 = 0.$$

Now the solution to be constructed approaches uniformity, i.e., $du/dx \to 0$ and $dT/dx \to 0$, as $|x| \to \infty$. It follows from (37.1–3), if the subscripts 1 and 2 are used to denote the respective values approached as $x \to -\infty$ and $x \to +\infty$, that

$$(37.5) \qquad \begin{aligned} \rho_2 u_2 &= \rho_1 u_1 = m, \\ p_2 + \rho_2 u_2{}^2 &= p_1 + \rho_1 u_1{}^2 = mP, \\ I_{02} &= I_{01}. \end{aligned}$$

These are precisely the shock relations (36.1–3). Their meaning is fully clarified if it is noted that the approach to these limits can depend only on a nondimensional measure of x, and it is seen from (37.2–4) that this cannot differ significantly from the variable Reynolds number

$$(37.6) \qquad \xi = \rho_1 u_1 \sigma \int_0^x \mu^{-1} \, dx.$$

The limits denoted by subscripts must therefore be approached already as $\xi \to \pm\infty$.

In terms of ξ as independent variable, and for $\sigma = \frac{3}{4}$, equations (37.2, 3) read

(37.7) $$u + m^{-1}p - du/d\xi = P,$$

(37.8) $$I_0 - dI_0/d\xi = I_{01},$$

by (37.4), where the constants are $m = \rho_1 u_1$,

$$P = u_1(1 + \gamma^{-1}M_1^{-2}) = (\gamma + 1)(u_1 + u_2)/(2\gamma),$$
$$I_{01} = u_1^2(\tfrac{1}{2} + (\gamma - 1)^{-1}M_1^{-2}) = (\gamma + 1)u_1u_2/(2\gamma - 2),$$

by (37.5), (36.7), (33.5), (35.8), and (36.4). But (37.8) implies

$$(I_0 - I_{01}) \exp(-\xi) = \text{const},$$

and (37.5, 6) show this constant to be zero, because $\mu > 0$, so that

(37.9) $$I_0(\xi) = \text{const}.$$

From (37.7), finally, by (37.1), (36.4), and (35.8),

$$u \, du/d\xi = u(u - P) + (\gamma - 1)(I_0 - u^2/2)/\gamma$$
$$= (\gamma + 1)(u_1 - u)(u_2 - u)/(2\gamma);$$

i.e.,

(37.10) $$\beta\xi = \frac{u_1}{u_1 - u_2} \log(u_1 - u) - \frac{u_2}{u_1 - u_2} \log(u - u_2) + \text{const}$$

with $\beta = (\gamma + 1)/(2\gamma)$. Together with (37.7) and (37.1), this is an explicit solution for ξ, p, and ρ in terms of u. Except at extreme pressures, experiment shows μ to be a smooth, increasing function of T for gases, and by (37.9), (35.8), and (36.4) μ is therefore a smooth, decreasing function of u^2 for the shock transition considered here and $x(u)$ is obtained from (36.6) and (37.10) by a quadrature. The corresponding curve of u versus x is sketched in Fig. 37.1.

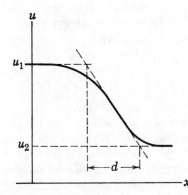

Fig. 37.1.

Observe that u, p, and ρ approach their asymptotic values *exponentially* as $\xi \to \pm\infty$, by (37.10), and hence also as $x \to \pm\infty$, because μ tends to a limit. when T does. An effective width of the transition zone may therefore be defined without significant ambiguity, for instance, by

$$d = (u_1 - u_2)/\max_{-\infty < x < \infty} |du/dx|$$

(Fig. 37.1). This *shock thickness* is given, according to (37.6) and (37.10), by

(37.11)
$$R_d\Sigma = 1,$$

where $R_d = \rho_1 u_1 d/\mu_1$ is the Reynolds number based on the shock thickness and the state approached as $x \to -\infty$, and

$$\Sigma = \max_{u_1 > u > u_2} \left| \frac{\sigma\beta\mu_1}{\mu} \frac{(u_1 - u)(u - u_2)}{u} \right| \Big/ (u_1 - u_2)$$

is a measure of shock strength, since $\Sigma/(M_1{}^2 - 1)$ tends to a nonzero limit as $M_1{}^2 \downarrow 1$. Hence, if L denotes a length scale characteristic of the observations, then

(37.12)
$$d/L \to 0 \quad \text{as} \quad R_L = \rho_1 u_1 L/\mu_1 \to \infty$$

for fixed, nonzero shock strength. The limit $R_L \to \infty$, with x/L and Σ fixed, is therefore one in which the shock transition does tend to a discontinuity, and it gives precise meaning to the limits used in Appendix 36. But as the shock strength $(M_1{}^2 - 1) \to 0$ for fixed R_L, however large, $d/L \to \infty$ and therefore a sufficiently weak shock is not thin.

For more general, Newtonian gases, the existence of a similar solution of (37.1–3) has been proved by qualitative methods of the theory of ordinary differential equations.

It is necessary, however, to close with the remark that gas dynamics does not furnish as adequate a basis for a theory of shocks as the foregoing might suggest. For fairly strong shocks (37.11) indicates a shock thickness d comparable to the mean free path l (Section 18). A fairly realistic, explicit solution is obtained for $T^{-\frac{1}{2}}\mu = \text{const}$, and this gives [49, Chap. 4]

$$
\begin{array}{ll}
d/l_1 = \infty & \text{for} \quad M_1 = u_1/a_1 = 1 \\
\quad\quad\ 4 & \quad\quad\quad\quad\quad\quad\quad\quad 2 \\
\quad\quad\ 2 & \quad\quad\quad\quad\quad\quad\quad\quad 5,
\end{array}
$$

where l_1 is the value of the mean free path corresponding to the state (ρ_1, S_1). Realization that this is difficult to reconcile with our premise that the gas maintains local thermodynamic equilibrium throughout the shock has led to considerable experimental and theoretical work. The results up to the time of writing indicate that the theory outlined here fits the facts much better than the values of d/l_1 suggest. One reason for this stems from the fact that the mean free path decreases in a strong shock transition to a final value l_2 much smaller than the initial value. The actual number of collisions in a shock is therefore considerably larger than d/l_1 indicates, and in a shock of moderate strength the departure from local equilibrium appears to be less serious than had been feared.

Problem 37.1. Show that for a perfect, monatomic Newtonian gas with $\sigma = \frac{3}{4}$ the entropy $S(x)$ has a *maximum* within the shock, whence it decreases (!) to the final value S_2. Why is such an entropy variation compatible with the Second Law?

Show also that the entropy assumes its maximum at the point in the shock where $u^2 = u_1 u_2$ (i.e., at the sonic point, by Problem 36.1).

38. Navier-Stokes Equations

It will be convenient for reference to collect here the differential equations governing the motion of compressible, Newtonian fluid. While the preceding sections have concentrated on gases, where the kinetic theory illuminates and guides the thermodynamical notions, the same considerations apply to Newtonian liquids in local thermodynamical equilibrium. Radiation will be ignored, as before. To convert the conservation principles into partial differential equations will require rather less than an appeal to the smoothness convention (Section 3); it suffices that the velocity and thermodynamic state are twice differentiable on the fluid domain. But the example of the preceding section, based directly on the conservation laws, has shown rather clearly why the Newtonian constitutive equation and Fourier's law may be expected to imply infinite differentiability, in any case.

Conservation of mass is expressed by (4.3),

$$(38.1) \qquad D\rho/Dt + \rho \operatorname{div} \mathbf{v} = 0,$$

where $D/Dt = \partial/\partial t + \mathbf{v} \cdot \operatorname{grad}$, from (3.1), and conservation of momentum, by (12.1),

$$(38.2) \qquad \rho \, Dv_i/Dt = \rho f_i + \partial p_{ij}/\partial x_j, \qquad i = 1, 2, 3,$$

where

$$(38.3) \qquad p_{ij} = (-p + \mu' \operatorname{div} \mathbf{v}) \, \delta_{ij} + 2\mu e_{ij},$$

from Postulate IXa; $2e_{ij} = \partial v_i/\partial x_j + \partial v_j/\partial x_i$. Thus if (38.2) be written out in full, it reads

$$(38.4) \quad \rho\left(\frac{\partial v_i}{\partial t} + v_j \frac{\partial v_i}{\partial x_j}\right) - \rho f_i = \frac{\partial}{\partial x_i}\left(-p + \mu' \frac{\partial v_k}{\partial x_k}\right) + \frac{\partial}{\partial x_j}\left[\mu\left(\frac{\partial v_i}{\partial x_j} + \frac{\partial v_j}{\partial x_i}\right)\right],$$

where the summation convention is understood, as always. On the assumption that radiation is absent, conservation of energy is expressed by the First Law (35.1), and it follows from (35.2), Corollary 4.1, and the Divergence theorem that

$$\int_{\Omega_t} \left\{ \rho v_i \frac{Dv_i}{Dt} + \rho \frac{DE}{Dt} - \rho v_i f_i - \frac{\partial}{\partial x_j} \left(v_i p_{ij} + \lambda \frac{\partial T}{\partial x_j} \right) \right\} dV = 0.$$

Since this holds for every fluid domain Ω_t (Section 2), and hence also for every open subset of such a domain, the integrand must vanish identically, and by (38.2),

$$(38.5) \qquad \rho \frac{DE}{Dt} = p_{ij} e_{ij} + \frac{\partial}{\partial x_i} \left(\lambda \frac{\partial T}{\partial x_i} \right),$$

because the symmetry of p_{ij} (Section 15) implies $p_{ij} \partial v_i / \partial x_j = p_{ij} e_{ij}$. In this form of the energy equation the work producing changes of kinetic energy has been split off to display explicitly the rate at which conduction of heat and work by the stresses produce changes of the thermodynamic internal energy per unit volume. Since the part $\rho p \, D(1/\rho)/Dt$ of this rate of work is stored reversibly in the fluid element (Appendix 35), and since (38.1) shows it to be equal to p div v, (38.5) is usually written in the form

$$(38.6) \qquad \rho \frac{DE}{Dt} = -\rho \operatorname{div} \mathbf{v} + \Phi + \frac{\partial}{\partial x_i} \left(\lambda \frac{\partial T}{\partial x_i} \right),$$

where, by (38.1), the *dissipation function*

$$\Phi = p_{ij} e_{ij} + p \operatorname{div} \mathbf{v}$$

represents the rate at which the stresses do work other than that which has already been identified as reversible. An alternative form of (38.6), by (35.7) and (38.1), is

$$\rho \, DI/Dt = Dp/Dt + \Phi + \partial(\lambda \, \partial T/\partial x_i)/\partial x_i,$$

and since $dI = c_p \, dT$ for a perfect gas [48] (even if the specific heats are not constant) the energy equation for such a gas may be written

$$(38.7) \qquad \rho c_p \frac{DT}{Dt} = \frac{Dp}{Dt} + \Phi + \frac{\partial}{\partial x_i} \left(\lambda \frac{\partial T}{\partial x_i} \right).$$

Equations (38.1), (38.4), and (38.7) are the system most commonly used as a basis for gas dynamical calculations. They are a system of five equations for the unknowns v_i, ρ, and T if p and c_p are expressed as functions of ρ and T by equations of state and μ, μ', and $\sigma = \mu c_p / \lambda$ are known functions of T.

(For liquids μ is generally a decreasing function of T, and σ is not necessarily constant; the information available on μ' is reviewed in [52].)

The dissipation function is, by (38.3), expressible as

$$\Phi = 2\mu e_{ij}e_{ij} + \mu'(\text{div } \mathbf{v})^2,$$

which shows it to represent the rate of work of the viscous stresses per unit volume. Since the *coefficient* $\mu' + 2\mu/3$ *of bulk viscosity*, like μ, is positive [18] (even though very small for perfect, monatomic gases), Φ is actually a positive definite form in e_{ij}, as may be seen by writing it out in components as

$$\Phi = 4\mu(e_{12}{}^2 + e_{23}{}^2 + e_{31}{}^2) + (\mu' + \tfrac{2}{3}\mu)(e_{11} + e_{22} + e_{33})^2$$
$$+ \tfrac{2}{3}\mu[(e_{11} - e_{22})^2 + (e_{22} - e_{33})^2 + (e_{33} - e_{11})^2].$$

This shows explicitly which part of the dissipation is due to the off-diagonal, or shear, stresses, and which to the diagonal, or normal, viscous stress components.† The matter is further illuminated by the Second Law (Appendix 35), according to which the rate of entropy increase of a body of fluid is

$$(38.8.) \qquad \frac{D}{Dt}\int_{\Omega_t} S\rho \, dV = \int_{\Omega_t} T^{-1}\left[\frac{DE}{Dt} + p\frac{D(1/\rho)}{Dt}\right]\rho \, dV,$$

by (4.4) and since the bracket is the rate of reversible heat addition to unit mass, by the First Law (Appendix 35). It follows from (38.1), (38.6), and the Divergence theorem that

$$(38.9) \quad \frac{D}{Dt}\int_{\Omega_t} S\rho \, dV = \int_{\Omega_t}\left(\frac{\Phi}{T} + \frac{\lambda}{T^2}|\text{grad } T|^2\right) dV + \int_{\partial\Omega_t} T^{-1}\lambda\mathbf{n}\cdot\text{grad } T \, dS.$$

and the last integral here represents clearly the rate of entropy increase due to addition of heat to the body of fluid, in its actual motion, by conduction across the boundary $\partial\Omega_t$. By contrast, the first integral on the right-hand side represents the rate of entropy increase due to the internal processes in the body of fluid, and its integrand is positive definite. The Second Law (Appendix 35) therefore shows that internal conduction of heat and dissipation of mechanical energy by the work of the viscous stresses are the two processes of irreversible heat addition in a Newtonian fluid. It is also worth noting the differential form of (38.9), which is

$$(38.10) \qquad \rho T\frac{DS}{Dt} = \Phi + \frac{\partial}{\partial x_i}\left(\lambda\frac{\partial T}{\partial x_i}\right),$$

by (38.1), (38.6), and Corollary 4.1.

---

† The Stokes relation (37.4) is adopted by most books on gas dynamics, regardless of the nature of the gas, and no appreciable error arises from this in the study of most compressible fluid motions because $(\mu' + 2\mu/3)|\text{div } \mathbf{v}|$ is usually too small to be significant. Exceptions are shocks, sound absorption, and acoustic streaming, where important viscous effects arise from diagonal stress components.

Problem 38.1.   The steady, axially symmetrical, compressible boundary layer equations on the forward portion of a body of revolution are

$$\frac{\partial(\rho u)}{\partial x} + \frac{\partial(\rho v)}{\partial y} + \frac{\rho u}{r_0}\frac{dr_0}{dx} = 0$$

$$\rho\left(u\frac{\partial u}{\partial x} + v\frac{\partial u}{\partial y}\right) = -\frac{dp}{dx} + \frac{\partial}{\partial y}\left(\mu\frac{\partial u}{\partial y}\right)$$

$$\rho\left(u\frac{\partial I}{\partial x} + v\frac{\partial I}{\partial y}\right) = u\frac{dp}{dx} + \frac{\partial}{\partial y}\left(\lambda\frac{\partial T}{\partial y}\right) + \mu\left(\frac{\partial u}{\partial u}\right)^2$$

where x is measured along the meridian of the body surface and y normal to it, and $r_0(x)$ denotes the radius of the body surface. Show that the rate of heat transfer to the gas across unit body surface area is

$$r_0^{-1}\frac{\partial}{\partial x}\int_0^\infty r_0\rho u\,(I_0 - I_{0e})\,dy.$$

Use Mises' transformation (Section 25) and Problem 4.2 to relate this boundary layer to a two-dimensional, compressible boundary layer.

Bibliography

[1] R. Courant, *Differential and Integral Calculus*, Interscience, New York, 1936.

[2] F. Brauer and J. Nohel, *Ordinary Differential Equations*, Benjamin, New York, 1966.

[3] O. D. Kellogg, *Foundations of Potential Theory*, Springer, Berlin, 1929.

[4] C-S. Yih, *Dynamics of Nonhomogeneous Fluids*, Macmillan, New York, 1965.

[5] M. Spivak, *Calculus on Manifolds*, Benjamin, New York, 1965.

[6] A. H. Wallace, *Algebraic Topology*, Pergamon Press, New York, 1957.
M. Greenberg, *Lectures on Algebraic Topology*, Benjamin, New York, 1967.

[7] L. M. Milne-Thomson, *Theoretical Hydrodynamics*, 2nd ed., Macmillan, London, 1949.

[8] A. Betz, *Konforme Abbildung*, 2nd ed., Springer, Heidelberg, 1966.

[9] H. Lamb, *Hydrodynamics*, 6th ed., Cambridge University Press, 1932, pp. 236–242.

[10] G. N. Ward, *Linearized Theory of Steady High-Speed Flow*, Cambridge University Press, 1955.

[11] G. K. Batchelor, *The Theory of Homogeneous Turbulence*, Cambridge University Press, 1956.

[12] N. I. Muskhelishvili, *Singular Integral Equations*, Noordhoff, Groningen, 1953.

[13] L. Prandtl, *The Essentials of Fluid Dynamics*, Blackie, London, 1952.

[14] S. Goldstein, *Lectures on Fluid Mechanics*, Interscience, New York, 1960.

[15] H. Grad, Statistical Mechanics, Thermodynamics, and Fluid Dynamics of Systems with an Arbitrary Number of Integrals, *Commun. Pure Appl. Math.* 5, 455–494 (1952).

[16] W. Prager, *Introduction to Mechanics of Continua*, Ginn, Boston, 1961.

[17] G. N. Patterson, *Molecular Flow of Gases*, Wiley, New York, 1956.

[18] S. Flugge (Ed.), *Thermodynamics of Gases, Encyclopedia of Physics*, Vol. 12, Springer, Heidelberg, 1958.

[19] S. Chapman and T. G. Cowling, *The Mathematical Theory of Non-uniform Gases*, Cambridge University Press, 1952.

[20] H. Jeffreys, The Wake in Fluid Flow Past a Solid, *Proc. Roy. Soc.* **A128**, 376–393 (1930).

[21] L. Rosenhead (Ed.), *Laminar Boundary Layers*, Clarendon Press, Oxford, 1963.

[22] H. Schlichting, *Boundary Layer Theory*, Pergamon Press, London, 1955.

[23] W. A. Coppel, On a Differential Equation of Boundary Layer Theory, *Phil. Trans. Roy. Soc.* **A253**, 101–136 (1960).

[24] K. O. Friedrichs, Asymptotic Phenomena in Mathematical Physics, *Bull. Am. Math. Soc.* **61**, 485–504 (1955).

[25] M. D. Van Dyke, *Perturbation Methods in Fluid Mechanics*, Academic Press, New York, 1964.
J. D. Cole, *Perturbation Methods in Applied Mathematics*, Blaisdell, New York, 1968.

[26] H. C. Levey, Singular Perturbations: A Model Equation, Aeron. Res. Lab., Dept. of Supply, Melbourne, Australia, Aerodynamics Note No. 157, 1957.

[27] E. T. Whittaker and G. N. Watson, *Modern Analysis*, Cambridge University Press, 1927.

[28] C. C. Lin, *The Theory of Hydrodynamic Stability*, Cambridge University Press, 1955.

[29] H. C. Levey, The Thickness of Cylindrical Shocks and the PLK Method, *Quart. Appl. Math.* **17**, 77–93 (1959).

[30] J. W. Miles, On the Generation of Surface Waves by Shear Flows, *J. Fluid Mech.* **3**, 185–204 (1957); **6**, 568–598 (1959); **13**, 433–448 (1962).

[31] W. R. Wasow, *Asymptotic Expansions for Ordinary Differential Equations*, Interscience, New York, 1965.

[32] W. Walter, *Differential- und Integral-Ungleichungen*, Springer, Heidelberg, 1964.

[33] S. Kaplun, The Role of Coordinate Systems in Boundary-Layer Theory, *J. Appl. Math. Phys.* **5**, 111–135 (1954).

[34] S. Goldstein, *Modern Developments in Fluid Dynamics*, Clarendon Press, Oxford, 1938.

[35] L. Howarth (Ed.), *Modern Developments in Fluid Dynamics, High-Speed Flow*, Vol. 2, Clarendon Press, Oxford, 1953, Chap. 11.

[36] K. Nickel, Einige Eigenschaften von Lösungen der Prandtlschen Grenz-schicht-Differentialgleichungen, *Arch. Rat. Mech. Anal.* **2**, 1–31 (1958).

[37] S. Kaplun, *Fluid Mechanics and Singular Perturbations*, Academic Press, New York, 1967.

[38] F. K. Moore (Ed.), *Theory of Laminar Flows*, Princeton University Press, 1964.

[39] J. Serrin, Asymptotic Behavior of Velocity Profiles in the Prandtl Boundary Layer Theory, *Proc. Roy. Soc.* **A299**, 491–507 (1967).

[40] W. Velte, Eine Anwendung des Nirenbergschen Maximumprinzips für parabolische Differentialgleichungen in der Grenzschichttheorie, *Arch. Rat. Mech. Anal.* **5**, 420–431 (1960).

[41] B. Thwaites, Approximate Calculation of the Laminar Boundary Layer, *Aeron. Quart.* **1**, 245–280 (1949).

[42] E. C. Maskell, Flow Separation in Three Dimensions, Roy. Aircraft Est., Min. of Supply, London, Rep. Aero. 2565, 1955. Also *Encyclopedia of Physics*, Vol. 8/1, Springer, Heidelberg, 1959, pp. 318–322.

[43] H. P. Greenspan, *The Theory of Rotating Fluids*, Cambridge University Press, 1968.

[44] H. W. Emmons (Ed.), *Fundamentals of Gas Dynamics*, Vol. 3, Part B, of *High Speed Aerodynamics and Jet Propulsion*, Princeton University Press, 1958.

[45] R. Courant and D. Hilbert, *Methods of Mathematical Physics*, Vol. 2, Interscience, New York, 1952, Chap. 2.

[46] S. Flugge and C. Truesdell (Eds.), *Fluid Dynamics III*, Encyclopedia of Physics, Vol. 9, Springer, Heidelberg, 1960, Chap. 3.

[47] R. Courant and K. O. Friedrichs, *Supersonic Flow and Shockwaves*, Interscience, New York, 1948.

[48] W. G. Vincenti and C. H. Kruger, *Introduction to Physical Gas Dynamics*, Wiley, New York, 1965.

[49] L. Howarth (Ed.), *Modern Developments in Fluid Dynamics, High-Speed Flow*, Vol. 1, Clarendon Press, Oxford, 1953, Chap. 2.

[50] R. C. Tolman, *Relativity, Thermodynamics and Cosmology*, Clarendon Press, Oxford, 1934, Chap. 5.1.

[51] W. D. Hayes and R. F. Probstein, *Hypersonic Flow Theory*, 2nd ed., Academic Press, New York, 1966.

[52] S. M. Karim and L. Rosenhead, The Second Coefficient of Viscosity of Liquids and Gases, *Rev. Mod. Phys.* 24, 108–116 (1952).

Index

Numbers in italics refer to the page where a notion is defined or introduced.

179